PERGAMON INTERNATIONAL LI
of Science, Technology, Engineering and Soci
The 1000-volume original paperback library in aid
industrial training and the enjoyment of leisure
Publisher: Robert Maxwell, M.C.

Thinking With Models

THE PERGAMON TEXTBOOK
INSPECTION COPY SERVICE

An inspection copy of any book published in the Pergamon International Library will gladly be
sent to academic staff without obligation for their consideration for course adoption or
recommendation. Copies may be retained for a period of 60 days from receipt and returned if not
suitable. When a particular title is adopted or recommended for adoption for class use and the
recommendation results in a sale of 12 or more copies, the inspection copy may be retained with
our compliments. The Publishers will be pleased to receive suggestions for revised editions and
new titles to be published in this important International Library.

International Series in
MODERN APPLIED MATHEMATICS AND COMPUTER SCIENCE

General Editor: **E. Y. RODIN**
VOLUME 2

NOTICE TO READERS

Dear Reader
If your library is not already a standing/continuation order customer to this series, may we recommend
that you place a standing/continuation order to receive immediately upon publication all new volumes.
Should you find that these volumes no longer serve your needs, your order can be cancelled at any time
without notice.

ROBERT MAXWELL
Publisher at Pergamon Press

Thinking With Models

Mathematical Models in the
Physical, Biological, and Social Sciences

by

THOMAS L. SAATY

and

JOYCE M. ALEXANDER

PERGAMON PRESS

OXFORD · NEW YORK · TORONTO · SYDNEY · PARIS · FRANKFURT.

U.K.	Pergamon Press Ltd., Headington Hill Hall. Oxford OX3 0BW, England
U.S.A.	Pergamon Press Inc., Maxwell House, Fairview Park, Elmsford, New York 10523, U.S.A.
CANADA	Pergamon Press Canada Ltd., Suite 104, 150 Consumers Road, Willowdale, Ontario M2J 1P9, Canada
AUSTRALIA	Pergamon Press (Aust.) Pty. Ltd., P.O. Box 544, Potts Poit, N.S.W. 2011, Australia
FRANCE	Pergamon Press SARL, 24 rue des Ecoles, 75240 Paris, Cedex 05, France
FEDERAL REPUBLIC OF GERMANY	Pergamon Press GmbH, 6242 Kronberg-Taunus Hammerweg 6, Federal Republic of Germany

First edition 1981

British Library Cataloguing in Publication Data
Saaty, Thomas Lorie
Thinking with models.—(International series in
modern applied mathematics and computer, vol. 2).
(Pergamon international library)
1. Title II. Alexander, Joyce M
III. Series
511'.8   QA402   80–41861

ISBN 0–08–026475–1 Hardcover
ISBN 0–08–026474–3 Flexicover

Printed in Great Britain by A. Wheaton & Co. Ltd., Exeter

Dedication

To all our co-workers in the field of modeling, particularly those whose models we did not have room to include.

Above all, we dedicate this work to PROFESSOR MANFRED ALTMAN, of the University of Pennsylvania, a champion of creative research, whose untimely death was a great loss to all of us.

Preface

In the past decade there has been a surge of interest in mathematical modeling. This is partly due to a demand for relevance of theoretical material in the real world and partly due to an intense search for methods to deal with complex societal problems. Many books have been written on the subject. The original material of this book dates back to the sixties. However, its spirit will probably be here for some time. It is to catalogue a wide variety of models for easy access and mental stimulation. It is the idea of a variety of models, rather than a complete description of how to model, that is the significant feature of this work. No one person has the time or incentive to study in detail every published model. But one can glean useful ideas from short summaries of models in different areas.

Real-life problems are messy, approximate, interconnected, demanding, and time consuming. It is not easy to solve a worthwhile one in a short exercise. Thus one often resorts to simple analogies to train the mind and prepare it for the variety existing in complex reality.

Creativity has been defined on many occasions as a process of selection from a broad combinatorial number of possibilities. This implies that one should obtain a wide exposure in order to become aware of variety and should develop good judgment in order to make a wise selection. Our purpose here is to provide this wide exposure in order to broaden the reader's combinatorial awareness. We give as comprehensive an account of mathematical models as possible in a book of this size. However, we must add that the improvement of judgment requires extensive training and experience; examples alone are insufficient for this purpose.

We have avoided philosophizing about the "wherefores" and "hows" and have concentrated our effort on examples of applications from several areas of human endeavor. Our purpose is not "to solve" but "to formulate," since there is a definite need to improve the ability to structure and formulate problems creatively. "Solving" is an established art in our colleges, universities, and scientific laboratories. "Formulation" is still a mystery relegated to the not yet well understood process of creativity. We feel that formulation can also be taught successfully at school in spite of the lack of real understanding of its methods and practices. When approached pragmatically, formulation of problems as mathematical models must be done frequently, in a variety of ways, daringly, and without fear of making mistakes. In time, the modeler learns to improve his style and to increase his ability to select the type of model he needs. The basic principle is that exposure and practice have no better substitute in the business of modeling. This book is a step in the direction of providing examples. It is a synthesis of a variety of formulations which cuts across the broad terrain of mathematical tools.

In the past quarter of a century, operations research, systems analysis, management science, and other related fields have brought the power of scientific analysis into a variety of military, governmental, industrial, and management problems. These are problems whose underlying motivation is the decision process. Most of the successes have come from applications made to processes treated as physical systems, be they economic, dietary, product mix, or whatever.

In the physical sciences it has generally been possible to introduce measurement into a problem, thus enabling one to quantify and to model. From these models, theories were developed to explain the underlying causal relationships. As the complexity and variety of problems increased, computer simulation became a popular device for experimentation with solutions. However, this type of solution does not usually provide an adequate understanding of the problem that can be communicated easily. We remain hopeful that new mathematical and scientific theories (such as the analytic hierarchy process on which we have been working) will arise to help us in these new behavioral areas in which human factors are prominent.

We feel strongly that wide exposure to, and continued awareness of, the diversity of mathematical methods available for application to problems can help increase the problem-solver's range of choice among methods to formulate problems. The art of problem solving has been explored by a number of people, the foremost among whom is George Polya, whose *How to Solve It* is universally known and has been translated into many languages. In his writings, Polya confines his attention to solving mathematical problems with mathematical tools. Our purpose is much broader as we attempt to reach the limits of the natural and social sciences. We should point out, however, that many of the models in the biological and social sciences are initial attempts to develop explanatory models and are not necessarily valid.

The book was conceived and developed in several drafts over the years by the first author. The earlier drafts have been used in courses at the University of Pennsylvania and in the National Science Foundation Chautauqua-Type Short Courses for College Teachers which he gave for the American Association for the Advancement of Science. The collection has been constantly updated and reduced in size so as to provide a reasonable-length book. In this last draft, a short section of models based on the theory of analytic hierarchies and priorities, developed by the first author, is also included. Some of the earlier models are still the best ones. Other models are in such a state of constant change that it was best to stick with one version of them; still we believe that this work represents the state of the art in modeling.

The material can be used for teaching purposes in a semester course on modeling or as supplementary reading for a two-semester course. It can also serve as a useful reference book for practitioners in the natural sciences and in operations research and management science.

The asterisk (*) indicates those sections that may be difficult or require special knowledge but may be skipped without affecting the understanding of the other material.

We hope that this book will be a learning tool for teachers as well as for students; we also hope to learn ourselves from those who use the book. To this end we ask that teachers who have discovered good exercises they would like to see in the book send them to us to be included in the next revision with full acknowledgments to them. We also welcome all comments and suggestions. Since modeling is an ever-growing field, we wish to encourage the idea of learning from each other to maintain the forward thrust of the field.

Contents

Part I.   Methodology

Part II.   Framework

PART I

METHODOLOGY

In the first two chapters some philosophy and methods of modeling are described. In the first chapter we give a broad view of modeling with a discussion of some of its objectives, advantages, and requirements. In the second chapter, methods of counting, estimation, and structuring—essential for the elementary approach to model building—are illustrated.

Chapter 1

An Overview

1.1. Introduction

Mathematical creativity consists in either recognizing that an existing formalism is applicable to the problem at hand or inventing a new one. [Kac and Ulam.]

SOLVING problems involves the use of two types of talent: imagination and skill. In general, the classroom approach to problems tends to emphasize skill because, to a certain extent, skill can be communicated through the knowledge of techniques. However, it is also possible to exercise the imagination through exposure to the analysis of a diversity of imaginatively formulated problems. Formulation seems to depend on the ability (and the knowledge) of the individual to structure problems. The pursuit of solutions is well documented in the mathematical and scientific literature. Considerably much less formal knowledge is available regarding formulation than there is about solution. The only way we have so far to train an individual in modeling is to expose him to a wide variety of problems and to a corresponding variety of models which provide representations of those problems. This establishes the need for a methodological framework for problem formulation.

We should make it clear that this chapter is not intended as a summary of what is covered in the rest of the book. It is designed as our introduction to modeling with some ideas about its philosophy and practice. The material here will be helpful to the student, since it will give him some general ideas about modeling and since it provides a framework or guide in helping him to structure problems. It offers him criteria for testing not only the models in this book to see how they meet the needs of his problems but also his own models as they develop. This chapter is designed to show the variety of thinking about modeling in general, just as the rest of the book is designed to show a variety of models; this is the spirit in which the chapter should be viewed.

1.2. Problems: Identification, Formulation, and Solution

When faced with a problem one first attempts to brainstorm all its relevant aspects, its internal variables, its parameters, and interactions with external factors. Some of the results of brainstorming are qualitative, others are quantitative. The next stage involves a classification of the results of the brainstorming into groups of ideas which belong together. Modeling involves a careful examination of the quantitative components to determine what can be measured and what cannot. The measures of effectiveness of the process under study must be stated and incorporated into a model. In grocery-store

3

queues, for example, the measure of effectiveness is the number of people able to pass through the line per unit of time. These measures define the problem by showing the degree of fulfillment which is possible with the current system and indicating what would be preferred. Often a statistical approach to a scientific problem is the first that comes to mind. Then an estimation model, often formed from a hybrid of statistics and algebra, is used to find a rough way of pursuing the answer.

The problem may be deterministic or it may be subject to chance occurrences and therefore probabilistic. In general a deterministic problem can have a descriptive or a normative setting. In a descriptive framework, equations and inequalities are used to relate the variables of the problem; these equations or inequalities may be algebraic, differential, difference, or integral. One or many solutions of a descriptive framework may be possible. In the normative setting of a problem, there is usually an objective function to be maximized or minimized, subject to descriptive equations or inequalities as constraints. Here one seeks out the optimal from among many possible solutions.

If a problem involves both optimization and probabilities, then the approach would ordinarily require maximizing or minimizing expected utilities after first defining the utility or objective function in terms of the random variables of the problem.

The qualitative description of a problem involves the definition of the objectives being pursued. This is followed by an account of the real-life process to be adopted to attain the objectives. It is essential to identify both controllable and uncontrollable factors in the process. A supermarket can control the flow of goods during its working hours and the number and efficiency of its checkout clerks. However, it has no major control over what a customer may want to buy, although it may control this somewhat by making only certain types of food available.

Analysis of the process requires gathering data and other information. In order to solve a problem it must be decided at the outset how the study is to be conducted. The conclusions and the way they are to be communicated must be developed for persuasion or selling purposes according to whether one is serving as consultant, lawyer, consumer, sponsor, law-enforcer, or decision-maker. Hypotheses are constructed on the basis of the information available, and a model is formed to test the hypotheses. The problem is then analyzed quantitatively through a suitable model which may be based on geometry, statistical analysis and correlation, ratio and proportion, rate of change, input/output (or matrix relations in general), linear algebra, equivalence, ordering preference with weight assignment, probability, optimization with single interest (programming, variational, or control theoretic) or multiple interests (game theory), and graphs and discrete mathematics. In any case, the model may be a first cut at the problem in the form of an estimation of the answer, as an upper or lower bound. Some useful guidelines about problems and problem solving will now be considered.

1.2.1. GUIDELINES ON FORMULATING THE PROBLEM

1. Define the problem and give its history and its causes.
2. State the objectives and the constraints.
3. Are you sure that this is the problem you want to solve? Why do you want to solve the problem?
4. Is there anyone who needs the solution? Are you sure you need the solution? Why?

5. Are there other related problems, perhaps easier, which should be solved first? List them.
6. What is the solution needed for?
7. What effect would it have?
8. How would you implement it?
9. How much would it cost to solve the problem? What are the resources available?
10. How much benefit would there be from the solution?
11. If the problem is ignored, will it terminate over time?
12. Get outside the problem and look at it. Is it significant?
    What is your vantage point for this judgment?
13. What are the stable solutions of the problem?
14. Can change in law or administration eliminate the problem?
15. Can the problem also be viewed as someone else's problem?
    Perhaps you can engage his cooperation in modeling a solution?

1.2.2. GUIDELINES ON HOW TO SOLVE THE PROBLEM

1. Does the problem have a solution?
2. Give all alternative solutions: are they exhaustive? How do you demonstrate this?
3. Give an optimal or near-optimal solution.
4. Give an average solution.
5. Give an approximate solution.
6. Start at both ends: the raw data and a hypothesized answer and move toward the middle to develop justification.
7. Start in the middle and move towards the ends.
8. Embed the problem in a larger context and solve it.
9. Abstract the problem to a simpler formulation.
10. Can you derive the solution from that of a related problem?
11. Simulate the problem in search of solution.
12. Construct a working hypothesis.
13. Develop and test the hypothesis.
14. Define the utilities and the payoffs in the process being studied.
15. How sensitive is the solution to changes in the data?
16. What are the invariants of the problem as reflected in the solution?
17. Update feedback of the implemented solution onto the problem under study.
18. Analyze the faulty solutions of the problem to get a better understanding of the preferred ones.

1.2.3. IMPACT OF THE SOLUTION

1. How can you communicate the problem and its solution to others? What is the most effective way to convince different people of your solutions?
2. What is the impact of the solution on people, things, etc.?
3. What people should be involved in implementing the solution?
4. What personnel commitments, organizational structure, and equipment are needed to find the solution and, in particular, to implement it?

5. What happens to the organization after the problem has been solved and the solution implemented?
6. Can the organization solve other problems?
7. How should people be motivated to solve the problem? .
8. What are the sanctions on, and threats to, the individuals and organizations involved?
9. What is the moral impact of the problem and its solution on people?
10. Will there be a chain reaction, either creating new problems or solving other problems, as a result of this solution?
11. Are there residual problems which must now be solved?

Knowing that the foregoing comprises major steps in problem solving, take a problem and brainstorm its description and solution with a group. Attempt to quantify the problem. Indicate the type of information needed to make the solution operationally useful.

1.3.   Mathematical Models and their Applications

The underlying idea in most of the models presented here is to indicate, in the form of an equation or inequality, (i) an algebraic relationship between variables, (ii) the rate of change of some variables with respect to other variables, and (iii) the sums or integrals of functions in order to obtain cumulative values and to see how they relate to other variables or functions. Naturally, some models differ from these basic types, but the majority will fall into these three classes. The fields of optimization and of stochastic processes each utilize these classes of formulations within a particular framework which is derived from concepts and operations peculiar to their fields. In studying formulations of models it is well to bear in mind that the number of basic ways available for formulation is very limited. The challenge is to ascertain that the given formulation is appropriate to the solution of the problem under study.

Roughly speaking, mathematical models may be divided into two types:

(i) Quantitative or based on the number system. By means of this category of models one attempts to answer questions asking "How many?" or "How much?" It can be used to express relations between elements and properties of systems.

(ii) Qualitative, possibly based on set theory, but not reducible to numbers. With this type of model, one studies relations between systems and their properties. Frequently the formulation of a quantitative model is preceded by a qualitative analysis of the problem under study.

Qualitative mathematical models include the use of axiomatics, set theory, group theory, and graph theory. Quantitative models may in turn be divided into two general categories, continuum (relating to the real or complex number fields), and discrete (relating to the integers). Discrete mathematics frequently involves counting and estimating numbers. Its major branches include Diophantine equations and number theory, optimization in integers and mixed optimization, combinatorial mathematics, certain aspects of graph theory, and geometric number theory. For our purpose, continuum mathematics may be taken as nondiscrete mathematics.

Both categories involve concepts from three major fields: (i) equations and inequalities,

(ii) optimization, and (iii) probability and stochastic processes. They can be displayed, as in the following triangle (Fig. 1.1).

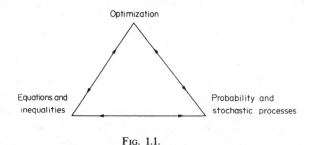

FIG. 1.1.

The abstract study of these three fields rests mostly in the realm of functional analysis. There is interaction between each pair of these three areas. For example, the field of optimization involves equations and inequalities which describe constraints on the system. Stochastic processes also use equations and inequalities to describe the behavior of systems. Optimization problems are increasingly recognized as occurring in conjunction with probability.

It is not an oversimplification to say that both major types of mathematics, qualitative and quantitative, are concerned with solving problems whether by characterization or by construction. This process may be outlined as follows.

1.3.1. THE PROCESS OF SOLVING

1. *A priori bounds.* One often begins by establishing the maximum number of solutions a problem can have. If in this process none is found to exist, there is no need to proceed further. If there are several one may decide to pursue a most desirable one in some sense. There is a saying, "There are many ways to skin a cat." But there is only one way to satisfy thirst, for example, and there are no ways for running up Mount Everest from the bottom in one hour.

2. *Existence and uniqueness.* Are there any solutions? This differs from the previous step in that one must prove that there is a solution instead of simply putting a bound on the number. Also, there may be no solution. For example, consider the problem of covering a chessboard from which two diagonally opposite squares have been removed. Is it possible to cover it with dominoes, each covering exactly two squares without overlapping? It may easily be seen that there are no solutions to this problem. (Consider the colors of the missing squares and the parity of covering by a single domino.)

3. *Convergence.* If an iterative instead of a closed form solution method is used, do the iterations converge to the desired solution?

4. *Approximations and errors.* An iterative process must be stopped. How good an approximation to the solution is given by the last iteration? What is the error incurred by this approximation?

1.3.2. EQUILIBRIUM AND STABILITY

The concept of equilibrium plays a central role in all important modeling whether relating to the solution of an equation, to an optimization problem, or to a stochastic process. We encounter the idea of equilibrium either directly or indirectly depending on the mathematical framework chosen.

It has been well recognized in the development of scientific theories that, to construct a useful model, one must adopt notions around which it is possible to analyze equilibrium and stability. These two ideas are pivots on which analysis revolves. A system is said to be in stable equilibrium if after a small disturbance it tends to return to its original state. It is unstable when a small disturbance tends to move it further and further away from its original state. A ball at the bottom of a salad bowl is in stable equilibrium, but a ball blocked by a pebble on a sloping street is unstable as a small disturbance will release it, never to return. Of course, oscillation is a form of stable behavior. In many social affairs stable equilibria are desirable situations and unstable equilibria are undesirable. For instance, an economic system violently fluctuating between boom and depression is undesirable, whereas one remaining in a well-balanced intermediate position is desirable. There have been several different approaches to political problems utilizing models based on stable equilibria. We will examine some later in this chapter.

There are instances in which stability is a bad thing and an unstable equilibrium is a good thing. To illustrate the first, consider the case of a person caught running between two tigers where the only way to escape is to get closer to one of the tigers. His momentary optimal position is halfway between. It is extremely dangerous for him to stay there, however, so he should make up his mind to escape in the direction of one of the tigers before they close in on him. He needs some extra psychological energy (that is, courage) to escape from this momentarily stable equilibrium.

An example illustrating the desirability of unstable equilibria is that of a small child, placed between two chocolate bars, who has been offered (and constrained) to take only one of them. For a moment he may be in a state of suspended equilibrium, unable to decide on which to choose, but he will soon escape from this unstable equilibrium situation of indecision by some minimal disturbance which attracts his attention to one of them and which makes him rush to that one. More generally, choosing between two good things is usually easy (unstable equilibrium), but choosing between two evils may be very difficult even if indecision is disastrous (undesirable stable equilibrium).

A further generalization of equilibrium is used when analyzing a conflict or competition situation involving several parties. In game theory the concept of "equilibrium point" refers to a situation in which none of the parties has an incentive to change his current strategy as long as the other parties do not change their strategies; the strategies at such a point are called "equilibrium strategies." It is also possible to classify equilibrium points according to whether they have short-range or long-range characteristics.

The conditions for equilibrium in the case of differential equations are generally obtained by equating to zero the velocity components of the system. The case where the coefficients are also functions of time is a more difficult one and gives rise to the idea of "dynamic equilibrium," which is a state of equilibrium in a continually changing system. The mode of approaching the equilibrium depends on the roots of the characteristic equation of the system. It is this analysis of the existence and character of the equilibrium that provides understanding of the long-run behavior of a system.

An example of "static equilibrium" is illustrated by a cold car engine which is slow to get started and which takes some time to warm up. During this time it vibrates, coughs, and is exceedingly reluctant to go. After a while this type of activity levels off as the engine warms up. One would never guess now that the car had been so unresponsive in leaving its state of inertia, which is one type of static equilibrium, for another state of static equilibrium, the steady state of operation.

An example of dynamic equilibrium is illustrated by a rigid rod, held at one end by a vertically oscillating pivot. Under some circumstances, the rod will remain upright so long as the pivot continues to oscillate and will fall down and behave like an ordinary pendulum once the oscillations cease. The stability of the system depends on the dynamics (time-varying behavior) of the pivot. Now for a comment on political equilibria.

Some have argued against equilibrium as a goal, saying that nations seeks to resist changes or disturbances, thereby hindering beneficial development. However, establishing a power equilibrium within and between nations, for example, may release in a country resources which would enable the pursuit of economic change. Thus, equilibrium need not be interpreted as an ultimate goal, but rather as a pause to integrate energies and resources in new directions, which may lead away from the initial equilibrium and serve as an incentive toward a new equilibrium. Essentially, we have a path up a mountain with resting places. The rests enable us to walk a steeper route than we could travel continuously.

A useful idea in mathematics is that of a property which remains invariant to change, and hence has the general characteristics of "stability." It is in this connection that mathematicians have used the idea of a fixed point, which is a convenient idea in modeling. The fixed point has its origins in topology and is amenable to wide interpretation when sets are transformed into themselves or into other sets. If T is a transformation mapping a set X into X, then a solution x_0 of the equation $Tx = x$, for x belonging to X, is known as a fixed point.

This idea is being used with a measure of success in looking for invariant properties in the social sciences.

1.4. The Modeling Process

There are many ideas to consider in the analysis and modeling of a problem. For example, a problem needs to be identified and background research is necessary before it becomes apparent what the real trouble is. Statistics and other types of information from people involved must be collected. An early estimate should be made of the resources required to study the problem; this involves both financial and problem-solving resources. A cost/benefit analysis might be conducted to determine the tradeoff between investing resources to solve the problem and leaving it alone because its effects are not as costly as its solution would be.

Once one has decided to study the problem, some method of specification or modeling process must be found. Initially, one may rely on the unwritten but widely used scientific principle that any model, no matter how crude, is to be preferred to no model at all. The idea here is that it is generally better to have someone think about a new phenomenon, even if the approach has obvious imperfections from the start. Those who come after can then improve on the simple model. Whatever model one may adopt, the question as to whether a solution to the problem is possible through the use of that model must be answered. People have been known to fall in love with an early formulation of a problem, regardless of whether or not it is valid for obtaining an answer.

In the process of formulation, dependent and independent, random and deterministic, variables need to be distinguished and identified and, as mentioned earlier, an analysis made as to which ones are controllable and which ones are uncontrollable. The parameters of the problem, together with their functional form as they appear in the model and whether they are deterministic or stochastic, must also be investigated. Sometimes, by developing a simple estimation model, a clear plan of attack crystallizes. If one is pressed for an answer, a "first-cut" model, sometimes called a "quick and dirty" model, may be developed. A more sophisticated model may need greater time.

An analytical model, as discussed previously, whether deterministic or stochastic, may not be possible. In that case, a simulation approach may be used instead, sometimes with the aid of a computer. Simulation has become increasingly popular recently. It is often necessary, but sometimes it is used at greater expense where a simple closed form model is available.

The use of ratio and proportion, and of rate of change, is central in setting up relations between variables to form equations and inequalities. After specifying the model and estimating its parameters, criteria are required to verify that the model is a reasonable representation of the problem. This may be tested by using previously collected data to validate the results.

A model should be simple enough to allow data collection and analysis. It should be practical in the sense that it might serve as an aid to implement the solution. Since prediction is focal in most scientific pursuits, most models must be designed with this point in mind. The generality of a model may be an enrichment on a special model developed for a particular problem. However, for the purpose of obtaining a solution, an approach to a problem may be decomposed into smaller models which are then aggregated for an overall result. A model, based on an efficient algorithm that can be used to derive the solution (hopefully without great cost), is preferable to another model from which a solution can only be obtained with great difficulty and considerable compromise and approximation. Algorithms are concerned with those methods of construction for which a list of instructions may specify a sequence of operations that, when followed, lead to the solution in a finite number of repetitions.

When a solution has been obtained, the question of the method and feasibility of implementation must be faced. The structure of the organization involved in this process and any means of making its operation more effective must be considered. The solution of a real-life problem does not stop with the theoretical solution but must be pursued to the point where the recommendations have actually been followed.

We must distinguish between two levels in modeling. There is the single level of penetration model which is short and to the point. There are two types: (a) those utilizing simple tools involving common sense, and (b) those using sophisticated tools. There is also a compound-level model which involves several strata of structure.

One can think of a number of reasons for using mathematical models. For example, models permit abstraction based on logical formulations using a convenient language expressed in shorthand notation, thus enabling one to visualize better the main elements of a problem and, at the same time, satisfying communication, decreasing ambiguity, and improving the chances of agreement on the results. A model allows one to keep track of a line of thought, focusing attention on the important parts of the problem. Models help one to generalize or to apply the results to problems in other areas. They also provide an opportunity to consider all the possibilities, evaluating alternatives, and eliminating the impossible ones. In science, in addition to prediction, they are a tool for understanding the real world and discovering natural laws. Today they are needed to cope with the various complexities of modern life, assisting and improving the ancient art of decision making. In the social sciences, a model can be pivotal in getting people to divulge their value judgments and discuss their view of a problem. In politics, the use of models has been known to provide opportunity to articulate individual feelings about how to approach a problem when such people were not willing or able to do so in other settings. The logic of the model transcends emotion and gets to the heart of the problem.

Suppose we are given a scientific problem which we wish to analyze for greater understanding of a situation and for prediction purposes. Assume that we are successful in formulating a mathematical structure with which we wish to portray faithfully the elements and relations between them in the problem under study. Since the mathematical parts and relations must correspond to the concepts of the problem, we speak of an isomorphism between the model and the problem. Thus, the model and its manipulations are a convenient logical tool for studying the behavior and relation of the elements of the problem.

There are two ways of looking at mathematical models—the local view and the global view, the micro and the macro, as the economists state it, or models which portray processes in the small and models which portray them in the large, as some mathematicians would think of it. It is generally believed that some experience with local modeling is indispensable in giving one a better perspective from which to approach global modeling.

Examples of powerful global theories that are inoperable on a local level are probabilistic models, such as statistical mechanics and queueing theory, which use averages and deviations by studying ensembles of objects and deterministic models of geodesic problems in the calculus of variations, algebraic topology, relativity theory, macroeconomics, and so on.

Local theories may also be very powerful ones; examples are point set topology, network flows with sources and sinks, and the analysis of the local shape of a surface (e.g., surfaces that are locally Euclidean, locally convex, or locally connected). In science, the study of a part of a large structure may enable one to gain understanding of its behavior and the way in which it might influence the whole structure.

Modeling involves a heroic simplification of a problem using the minimum possible number of basic variables in order to come to grips with the essentials. The first attempt usually comprises a stepping stone to more sophisticated elaborations of the model. To build an edifice, one requires a well-planned foundation, without which there could be no sound structure.

The process of model building is painstaking and experimental, involving hypothesis making, trial and error, and considerable innovative daring. In the early stages of modeling, one cannot afford to be too philosophically detailed, to demand a comprehensive statement of the alternative ways the problem can be stated, or to expect that each formulation make sense in all frameworks. The result might be to make it hopeless for structuring. Frequent experience with modeling through class participation indicates that the brainstorming process followed by a first cut at a model can be accomplished on some relatively complicated problem in about one and one-half hours. The results are psychologically impressive. The desired model may require much harder work, but the level of penetration obtained through such participation, and the courage derived from seeing this relatively crude, but fertile, flow of creativity, leaves one with the confidence that he can do the same on his own. Many have gone on to tackle and solve problems which have been discussed in this way, having benefited greatly from the experience.

Mathematical models have been remarkably effective in the physical sciences. However, they have fallen short in providing adequate representations of significant social science problems. Lack of criteria for the measurability of behavioral variables makes the introduction of relevant mathematical concepts difficult in the social sciences. But there have been some interesting breakthroughs leading to applications in areas heretofore regarded as beyond the reach of numbers. (See Chapter 8)

In the social sciences, modeling has utilized continuous mathematics for the following types of models:

1. A model based on the real numbers, money, length, mass, time, or some other metric as a cardinal scale. We call this type a quantitative-oriented model in the strict sense.
2. A stochastic or probabilistic model.
3. A stability model (frequently useful with ordinal types of utility) whose equilibrium may be either algebraic or geometric.
4. A surrogate model which is designed to portray the interactions of various factors in the absence of a useful way of measurement to obtain numerical answers. A "cardinal scale" is a measure of magnitude which can be found by counting units of measurement. In modeling applications, a "utility function" often serves as a substitute for a metric.

A model incorporates two ideas, the measurement of variables and parameters, and the relationship between these variables.

In a surrogate model, the relationship is displayed but without measurement. Hence, a surrogate model is incomplete. It is difficult to test the relationships it portrays, because means of measurement in this case are not dependable. On the other hand, some people have been able to prove that the solution of a surrogate model remains stable in spite of changes (transformations) of the relations.

Experience with a variety of models applied to conflict resolution has shown that ordinal utility models are more effective in practice than those employing cardinal utility simply because it is easier to apply ordinal scales of measurement to such problems. In the case of equilibria, the judgment of the parties involved in the conflict is used to generate the ordinal utility scale upon which the stability of policies (in a game theoretic framework) is analyzed.

Modeling is frequently used as an aid to decision making, i.e., in making choices among alternatives with the purpose of exercising control, subject to uncertainty in the system. Examples of criteria applied in making a decision are:

  (i) obtaining optimal performance;
 (ii) developing priorities prescribed by outside constraints;
(iii) humanization by increasing interaction between man and system, leading to learning and feedback to improve performance.

A number of models presented here have their origin in the science of decision making although their number is small compared to the variety of areas represented to which models have been applied.

1.5. Why structuring and solving should be separated

In this book, we focus on formulating problems and very rarely solve them because problem solving requires such considerable skill that it tends to interfere with the free, imaginative style involved in formulation. Too much problem solving creates a spirit of attachment to the particular method used because of the great investment of time and effort in the method. Granted that we formulate problems in order to solve them. However, we often formulate problems in order to decide which one we want to solve. Extensive experience with formulation serves the latter purpose well. Thus, our desire for breadth of scope prohibits giving solutions except in some simple cases.

One outgrowth of the direct attempt to structure a problem is to systematize the division of real-life problems according to their conditions, variables, and parameters, by a classification scheme which simplifies the choice of models. Frequently the choice of a model depends on the experience and taste of the problem-solver, rather than on properties inherent in the problem, though the latter should be the more valid restriction on the choice of a model.

Exposure to diverse models may induce some to wonder whether modeling is the art of contriving purely from the imagination, or whether models are actually dictated by the characteristics of the processes being modeled. For example, the answer to:

> Does the external world dictate the choice of axioms, definitions and problems or are these (models) in essence free creations of the human mind, perhaps influenced, or even determined, by its physiological structure? [Kac and Ulam.]

is, we believe, both.

An individual with a technical model needs the ability to persuade others to use it. This process may require time. People might be broken in on it piecemeal and without too much demand on their patience and good nature. The art of letting others see things your way is perhaps one of the most essential attributes of being effective [Boettinger]. Simply because one has an ingenious idea is no guarantee that others, even if they appreciate it, would take the trouble to use it. Effective modeling cannot be decoupled from human nature any more than the brain can live without the body. When viewed in this light the reader will quickly perceive why politics is a high-priority factor in decision making.

Chapter 1—Problems

The following problems are couched in very general terms and are designed to stimulate discussion. You are not expected to produce "answers" but should be able to formulate some ideas on the subjects. You may find it helpful to return to these questions at the end of your course to see if your answers would change.

1. Define the following terms: model, theory, hypothesis, conjecture, postulate, axiom, law, theorem, lemma, corollary, syllogism, deductive argument, inductive argument. Are your definitions adequate? Can you improve them?

2. Given the following two principles about proofs: (a) for someone to know a mathematical theorem he or she must check a proof of it; (b) if a proof of a mathematical statement exists, then, since a proof is rigorous, the statement is absolutely valid. What about a proof by computer which must work out many cases? The computer makes an error somewhere but, if it takes a long time to check an entire proof, is this a valid way of proof? Should we allow a degree of fallibility for proofs? Is $1 + 1$ always equal to 2? This sort of statement occurs often in life and one should be able to answer it in general.

3. Theories about nature usually grow and change. Does our understanding of natural law change? How about its applicability in experience: are the effects different?

4. It has been said that deductive thinking by going from cause to effect is linear thinking. How then is man ever to understand and to deal with complex problems which form a network rather than a chain? Is Aristotelean thinking the only way to understand the "real world" or are there other approaches? How effective are these? What are some criteria for evaluating progress with any method of understanding? What is meant by objectivity? Is there an objective reality? How do we know? Some have argued that objectivity is agreed upon subjectivity. What can this possibly mean?

5. In view of the foregoing, what can one say about the absoluteness versus relativeness of models from the standpoint of (a) people, (b) subject matter, (c) the time era in which the work is done, (d) historical accounts, (e) advocacy and politics, i.e., have good friends peddle your theories, (f) the fact that we see the world by the chemistry of our senses and our glands.

6. In view of 5, now comment about the objectivity of models.

7. Most mathematical models depend on basic "primordial" tools such as arithmetic, algebra, the calculus, and geometry. Have you felt a need for new basic mathematical ideas that could give a better expression for your experience and feelings? Can you describe the feelings and can you give a hint of what kind of new tool?

8. Clarify what is meant by understanding, prediction, control, and planning. Can we plan and modify our methods of understanding?

9. What can one possibly mean by evolving reality? Emergent reality? Do we make things up as we go along or do we discover things as they really are, awaiting us to find them?

10. The systems' way of thinking considers a number of different ramifications of a problem. Does this differ from the classical approach to modeling? What must we do to deal with problems more realistically?

References

Boettinger, Henry M., *Moving Mountains or The Art of Letting Others See Things Your Way*, Macmillan, New York, 1969.
Kac, Mark and Stanislaw M. Ulam, *Mathematics and Logic; Retrospect and Prospects*, A Mentor Book, The New American Library, New York, 1968.

Bibliography

Bartholomew, D. J., *Stochastic Models for Social Processes*, Wiley, New York, 1967.
Batschelet, E., *Introduction to Mathematics for Life Scientists*, Springer-Verlag, Berlin, 1973.
Beck, Anatole, Michael N. Bleicher, and Donald W. Crowe, *Excursions into Mathematics*, Worth, New York, 1970.
Beckner, Morton, *The Biological Way of Thought*, University of California Press, Berkeley and Los Angeles, 1968.
Bender, E. A., *An Introduction to Mathematical Modeling*, Wiley, New York, 1978.
Blalock, Hubert N., Jr., *Theory Construction: From Verbal to Mathematical Formulation*, Prentice-Hall, Englewood Cliffs, New Jersey, 1969.
Buffo, Elwood S. and Dyer, James S., *Model Formulation and Solutions Methods*, Wiley–Hamilton, New York, 1977.
Busacker, Robert G. and Thomas L. Saaty, *Finite Graphs and Networks*, McGraw-Hill, New York, 1965.
Churchman, C. West, *Thinking for Decisions: Deductive Quantitative Methods*, Science Research Associates, 1975.
Committee on Support of Research in the Mathematical Sciences, *The Mathematical Sciences*, MIT Press, Cambridge, Massachusetts 1969.
Coombs, C., *Mathematical Psychology: Elementary Introduction*, Prentice-Hall, Englewood Cliffs, New Jersey, 1970.
Emshoff, J. R. and Sisson, R. L., *Design and Use of Computer Simulation Models*, Macmillan, New York, 1970.
Gass, Saul I, Evaluation of complex models, *Computers and Operation Research*, Vol. 4, pp. 27–35, March 1977.
Gold, H. J., *Mathematical Modelling of Biological Systems: An Introductory Guidebook*, Wiley, New York, 1977.
Hitch, C., *Modelling Energy-economy Interactions*, Johns Hopkins University Press, 1978.
Lazarsfeld, Paul F. and Neil W. Henry, *Readings in Mathematical Social Science*, The MIT Press, Cambridge, Massachusetts, 1968.
Leik, R. K. and B. F. Meeker, *Mathematical Sociology*, Prentice-Hall, Englewood Cliffs, New Jersey, 1975.
Levin, S., *Some Mathematical Questions in Biology*, Mathematical Association of America, Part I: Cellular Behavior and the Development of Pattern, 1978; Part II: Populations and Communities, 1978.
Maki, Daniel P. and Maynard Thompson, *Mathematical Models and Applications*, Prentice-Hall, Englewood Cliffs, New Jersey, 1973.

Malkevitch, Joseph and Walter Meyer, *Graphs, Models and Finite Mathematics*, Prentice-Hall, Englewood Cliffs, New Jersey, 1974.

Miller, George A., *Mathematics and Psychology*, Wiley, New York, 1964.

Moder, Joseph J. and Salah E. Elmaghraby (eds.), *Handbook of Operations Research:* Vol I, *Foundations and Fundamentals,* Vol. II, *Models and Applications,* van Nostrand Reinhold, New York, 1978.

Polya, G., *Mathematics and Plausible Reasoning*, Vols. I and II, Princeton University Press, Princeton, New Jersey, 1954.

Polya, G., *How to Solve It*, Doubleday, Garden City, New York, 1957.

Rosen, R. (ed.), *Foundations of Mathematical Biology*, Academic Press, New York: Vol. I, *Subcellular Systems,* 1972; Vol. II, *Cellular Systems,* 1973; Vol III, *Supercellular Systems,* 1974.

Rubinow, S. I., *Introduction to Mathematical Biology*, Wiley, New York, 1975.

Saaty, Thomas L., *Mathematical Methods of Operations Research*, McGraw-Hill, New York, 1959.

Saaty, Thomas L., *Elements of Queueing Theory, with Applications*, McGraw-Hill, New York, 1961.

Saaty, Thomas L., *Mathematical Models of Arms Control and Disarmament*, Wiley, New York, 1968.

Saaty, Thomas L., and F. Joachim Weyl, *The Spirit and the Uses of the Mathematical Sciences*, McGraw-Hill, New York, 1969.

Saaty, Thomas L., *Optimization in Integers and Related Extremal Problems*, McGraw-Hill, New York, 1970.

Scientific American, *Mathematics in the Modern World*, W. H. Freeman, San Francisco, California, 1968.

Scientific American, *Mathematical Thinking in the Behavioral Sciences*, W. H. Freeman, San Francisco, New York, 1968.

Vemuri, V., *Modeling of Complex Systems*, Academic Press, New York, 1978.

Wall, R., *Introduction to Mathematical Linguistics*, Prentice-Hall, Englewood Cliffs, New Jersey, 1972.

Williams, H. P., *Model Building in Mathematical Programming*, Wiley, New York, 1978.

Chapter 2

Three Basic Themes:
Counting, Estimation, and Structuring

2.1. Introduction

THREE of the most powerful methods which play a prominent role in modeling are *counting, estimation*, and *structuring*. If a problem involves some finite set S, we often have to know exactly how many elements S contains, that is, we need to count the elements of S. One may count objects either directly through one-to-one correspondence or by means of relations which may exist between elements or sets. Estimation, on the other hand, is an art which improves with the amount of knowledge one has and which involves common sense based on a good use of the available facts. Estimation often serves as a basis for selecting models and orienting the direction of research: it may include the selection of reasonable upper and lower bounds or an acceptable range of values for the variable under consideration. We also include some examples of counting and estimation in a probabilistic setting. Structuring involves the seizing of some essential innate feature of the situation and then modeling the problem *with respect to* that feature. We close the chapter by looking at some common methods of proof which may be used in solving problems.

In Chapter 1 we discussed methods of modeling. We now turn our attention to results obtained using these methods.

We have already made brief mention of the type of outcome or result to be expected from a model. For example, we may obtain a number as an answer to a problem of "how many" or "how much;" we may find an upper or lower bound on a number; we may also obtain a function or a structure from which the solution may be derived. In fact, the number of forms which an answer may take is limited and an individual interested in modeling should be familiar with all of them. In this chapter we illustrate a large number of such results or outcomes by pursuing them through models constructed for the purpose of solving a problem. If one is familiar with how results are derived from models by appropriate manipulation, it becomes less pressing to solve every problem which may be encountered in modeling. This is, of course, an oversimplification of the diversity of results obtained from models (one may argue that no two problems are quite the same); still it seems obvious that with care the teaching of modeling would benefit greatly in time saved by not solving every problem completely but by concentrating on familiarization with many different types of models.

In addition to counting and enumeration, and estimating upper and lower bounds, we also have problems which involve finding maxima and minima (Chapter 4), and computing probabilities, expected values, and deviations (Chapter 5). Numbers are also used to assign ranks to objects and to obtain numerical priorities or to establish a sequential order in which they should occur. Other common forms which mathematical models may take are algebraic structures: (i) equations (Chapter 3) which can be used to obtain unknown numbers or functions; (ii) inequalities which may be used to establish bounds; and (iii) optimization methods (Chapter 4) used to locate maxima and minima as numbers or functions. Functions are used to describe relations and may be analytical, geometric, or tabular in form. Examples of more abstract algebraic structures which have found their way into applications are groups and lattices.

Another common form which a mathematical model may take is a geometric structure. Many problems formulated in a geometric setting such as those in Euclidean space can be translated into an algebraic setting, but there are others that cannot. Among these are problems in graph theory, geometric number theory, discrete geometry, and the geometry of numbers. Both algebraic and geometric structures are used to analyze conditions for the stability or instability of equilibrium of a system and to study other aspects of its behavior such as periodic oscillations or interaction with other systems. When a problem cannot be solved in one "whole" step it can often, by relaxing the original assumptions, be decomposed to obtain partial solutions, i.e., the problem can be decomposed into components and each solved in a partial step. The whole step solution is then composed of these "partial" steps either by simply combining the steps in a static composition of the partial solutions or by carrying out transformations in the model from step to step in such a way as to leave the "partial"-step solutions invariant in a dynamic composition of these solutions.

These examples cover a wide range of "results" which constitute a basis for constructing models to improve understanding, to predict, and to solve problems. Many other types of results will be found throughout the book.

2.2. Counting

As we have already noted, we frequently need an organized method of counting: an answer to the questions How many? or How much?

We start with a simple problem.

Example 1. A Simple Inventory Problem

We have an initial stock of X units of an item with a deterioration rate per year of L per unit. We use D items for stock at the end of the year. How much is left at the end of n years?

At the beginning of the first year we have X units.

At the end of the first year we have

$$X(1-L)-D.$$

At the end of the second year we have

$$\{X(1-L)-D\}(1-L)-D = X(1-L)^2 - D(1-L) - D.$$

At the end of the third year we have

$$\{X(1-L)^2 - D(1-L) - D\}(1-L) - D$$
$$= X(1-L)^3 - D(1-L)^2 - D(1-L) - D,$$

and at the end of the nth year we have

$$X(1-L)^n - D(1-L)^{n-1} - D(1-L)^{n-2} - \cdots - D(1-L) - D$$
$$= X(1-L)^n - D\frac{1-(1-L)^n}{1-(1-L)}$$
$$= \left(X + \frac{D}{L}\right)(1-L)^n - \frac{D}{L}.$$

This is the desired amount of stock left after n years. This process must terminate for that value of n which makes the remainder, before removal of D, less than D, i.e. for the value of n which first makes our expression above negative.

For example, let $X = 1000$, $L = 0.2$, $D = 100$. Then the stock after n years is given by

$$\left(1000 + \frac{100}{0.2}\right)(1-0.2)^n - \frac{100}{0.2} = 1500(0.8)^n - 500.$$

It is equal to 700 for $n = 1$; 460 for $n = 2$; 268 for $n = 3$; and 114.4 for $n = 4$. For $n = 5$ deterioration reduces the stock to less than 100 items, so the process will terminate.

We now consider three examples in which we calculate bounds for a required number: these bounds may or may not be attained.

Example 2. The Number of Cherries in a Can

We are often faced with estimating the maximum number of objects contained in a container, particularly when we are designing the container and the possible arrangement of the objects in the container. The following is a simple illustration of this.

How many cherries each of radius r can be packed in a can of radius R and height h? This will vary with the packing, so we obtain upper and lower bounds and refine them as far as possible. For example, no more cherries than the ratio of the volume of the can to that of a cherry can be packed in the can. A finer upper bound is obtained by multiplying the volume of the can by 0.74, the greatest known density for packing spheres, and then dividing the result by the volume of a cherry. As a lower bound we may first see how many circles of radius r can be packed in a circle of radius R. We then divide the height of the can by $2r$; the integral part of this ratio gives us the maximum number of layers of cherries. The product of these two quantities gives a lower bound.

Example 3. Family Tree and Counting Trees

This example illustrates the "essential brotherhood" of man. Let H be your family tree and let n be the number of generations represented in H. How large must n be so that sufficiently far back in time two branches of the family tree meet in the same ancestor? The population of the world today is approximately 4.5×10^9. We assume that the population of the world n generations ago was not more than 4.5×10^9. Now one individual living today descended from two individuals. They each in turn descended from two other individuals, and n generations ago the tree would have had $2n$ individuals at that generation level. These can be distinct only if $2n \leqslant 4.5 \times 10^9$. Hence if $2n \geqslant 4.5 \times 10^9$ the individuals could not be distinct, and two branches would have the same ancestor. If

$2^n = 4.5 \times 10^9$, then n is approximately equal to 32. Thus an upper bound to the number of generations we need to go back to find a common ancestor is 32.

We conclude this section with our final upper bound example.

Example 4. Length of a Path Containing n Points

We are often concerned with minimizing the distance travelled through a number of places. There is a long standing problem in the field of operations research called the traveling salesman problem (yet unsolved) which has the same concern in that the salesman must visit n cities traveling the shortest distance. The exact answer to this problem as formulated here in the Euclidean plane is not known but here is a good bound.

Given any n ($\geqslant 2$) points in a unit square, there is a path through the n points which does not exceed $(2n)^{1/2} + 1.75$ in length. [Few.]

Consider the unit square $0 \leqslant x \leqslant 1, 0 \leqslant y \leqslant 1$, and let the coordinates of the n points be $(x_1, y_1) \ldots (x_n, y_n)$. Consider q horizontal lines $y = 0, 1/q, 2/q, \ldots, 1$, with q arbitrary, and draw from each of the n points a perpendicular to the nearest of the $q + 1$ lines (Fig. 2.1). Repeat the construction, using the q lines, $y = 1/2q, 3/2q, \ldots, (2q-1)/2q$.

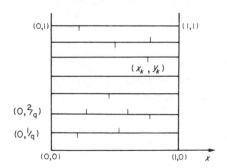

Fig. 2.1.

Each construction gives a path consisting of suitable portions of the lines, $x = 0$, $0 \leqslant y \leqslant 1$, and $x = 1, 0 \leqslant y \leqslant 1$, together with each perpendicular line counted twice— once from a horizontal line to a point and then back to the horizontal line—thus continuing the path through the n points. If we denote these two paths by L_1 and L_2, respectively, we have

$$L_1 = q + 1 + 2 \sum_{i=1}^{n} q^{-1} \|qy_i\| + 1,$$

$$L_2 = q + 2 \sum_{i=1}^{n} q^{-1} \|qy_i - \tfrac{1}{2}\| + 1,$$

where $\|\alpha\|$ denotes the absolute difference between α and the nearest integer. Note that

$$\|\alpha\| + \|\alpha - \tfrac{1}{2}\| = \tfrac{1}{2}.$$

We have

$$L_1 + L_2 = 2q + 3 + 2q^{-1}\frac{n}{2} = 2q + 3 + nq^{-1}.$$

We choose the integer q to minimize this value: it is the integer nearest to $(n/2)^{1/2}$. Thus $n = 2(q+\theta)^2$, with $|\theta| \leqslant \frac{1}{2}$. Substituting for n, we have

$$L_1 + L_2 = 2q + 3 + 2(q+\theta)^2 q^{-1}$$

$$= 4q + 3 + 4\theta + \frac{2\theta^2}{q}$$

$$= 2\{2(q+\theta)\} + 3 + \frac{2\theta^2}{q}$$

$$\leqslant 2(2n)^{1/2} + \frac{7}{2}.$$

Consequently the length of one of the two paths does not exceed $(2n)^{1/2} + 1.75$.

2.3. Results of Estimation

We have considered exact counting and upper and lower bounds to values of interest to us. We now move on to estimation, where we need a very approximate idea of size. As we have already noted, we use "estimation" in its everyday rather than in its statistical sense.

We often want a very rough approximation to a result in order to determine feasibility and practicality. Frequently our concern is simply to obtain an order of magnitude.

The following three examples illustrate these ideas.

Example 1. Chess Strategies

How many strategies are open to each player in a game of chess?

How long would it take a computer to analyze a game?

We consider the middle game. In general, White and Black will each have about 30 to 40 legal moves available. Combining each legal move of White with each legal move of Black gives about 1000 possible combinations of a single move of each player. Two consecutive moves would lead to a million different positions, three moves to a billion, four moves to a trillion and so on, multiplying the number of possible variations by 1000 each time a move is made.

A good chess game between evenly matched players averages from 40 to 50 moves. Since each new move generates 1000 possible combinations and each one must be combined with the 1000 variations of the next move, it follows that 25 moves each would generate a total of 10^{75} variations. Assume that a computer can analyze a million variations per second; it would take the computer 10^{69} sec to decide which move to make next.

How long is 10^{69} sec? Well, our planetary system is estimated to be $4\frac{1}{2}$ billion years old, which in round numbers is 10^{18} sec.

This example leads us to the conclusion that a chess-playing computer should not try brute-force exhaustive analysis of the next 25, or even of the next 10, moves. Instead, it should probably follow the experts and depend upon positional analysis and tactical maneuvering, letting long-term strategy take care of itself.

The next example deals with astronomical distances, where we are often satisfied with order of magnitude approximations.

Example 2. The Diameter of the Observable Universe [Asimov]

We are concerned with estimating the size of the observable universe.

Hubble's law states that the velocity of recession of a galaxy varies directly with its distance from the earth. Thus, we may "reach" a point where a galaxy is receding from us at the speed of light and in consequence nothing from that galaxy or points beyond can be observed by us. The problem is to estimate this distance and hence the diameter of the observable universe.

By Hubble's law [1]

$$V = kD,$$

where V is the velocity of recession in miles/sec, D, is the distance from the earth in millions of light years, and k is Hubble's constant.

To find k, we use the following information. The Virgo cluster is receding from us at 710 miles/sec, as shown by the red shift of its components. The brightness of the Virgo cluster when compared with the brightness of the Andromeda galaxy shows that it is 16.5 times as far away as the Andromeda galaxy, which is 2,300,000 light years away.

Thus for Virgo,

$$D = \frac{16.5 \times 2,300,000}{1,000,000} = 38.$$

Therefore $k = 710/38 = 18.5$. Some recent calculations suggest that this is somewhat large. We assume that $k = 15$. Using $V = 186,282$ miles/sec (speed of light) gives

$$D = \frac{186,282}{15} = 12,500.$$

Consequently, the limit of observation is 12,500 million light years away, or the diameter of the observable universe is 25 billion light years.

The next example relates to information theory. In that field a bit is defined as a unit of computer memory corresponding to the ability to store the result of a choice between two equally probable alternatives.

Example 3. Brain Storage

Here we estimate the upper limit of the amount of information that the human mind can handle. An individual cannot give his full attention all the time, but if he could and if he were able to handle 20 bits of information per second (recall that this means $2^{20} \approx 1,000,000$ different possibilities per second) then over a 14-hour day he would handle 10^6 bits. Over 50 years this would become over 10^{10} bits or $2^{10^{10}}$ possibilities. Even so, he would not need the total capacity of his nervous system to store this information. Consider that there are 10^{10} neurons in the human brain and 10^{12} interneuron connections or synapses so that the number of possible networks which can be formed from the neurons and available synapses is a very large number. Only a small number of the available cells is necessary for storage.

The French physicist J. C. Levy has shown that only one millionth of the available cells would be required to store the amount of information we are discussing. We can also note that a man can retrieve 5×10^{10} bits of information per second whereas a computer can retrieve 10^9 bits/sec: the computer is almost as efficient as a man in this respect.

2.4. Counting and Estimation and their use in Probability

Counting may arise in at least two ways in problems where probability is involved. One may have to calculate the number of ways in which something can happen (as in classical probability), or one may need the expected value of a quantity.

We have already noted that estimation can be used at a preliminary stage in model building; it can also be used later to give a rough check on whether a model which has been derived is reasonable. We use it here in a test of a probability model.

Example 1. Passenger Pick-up and Discharge Capacity

We need to know if there is adequate parking space for 600 cars each hour to discharge passengers at the curb of a 1000-ft frontage of the enplaning level of a new Pan-American terminal. A car occupies a space for about 102 sec. For a number of reasons, it is possible to provide only 28 parking spaces along this frontage. It is assumed that the cars arrive by a single lane into three-lane traffic at the frontage and that the right lane is used for parking, the middle lane for maneuvering, and the left lane for movement.

If we assume that the left lane moves at 5 mph, while cars move into and out of it, and that each car is 17 ft long, and if we allow a car length between cars for each 10 mph, we must allow approximately 25 ft of space per car. In 1000 ft there is room for 40 cars in a single lane. At 5 mph it takes almost $2\frac{1}{2}$ min for a moving car to clear the frontage. Thus, in one hour, we can move through a single lane (without parking) at most $(60 \div 5/2) \times 40 = 960$ cars, which is more than the 600 needed.

Now the service rate of a space is $3600/102 \approx 35$ cars per hour. Since there are 28 spaces, then $28 \times 35 = 910$ cars per hour may park, which is well in excess of the peak hour requirement of 600 and below the 960 which can move through the area. Of course we are assuming ideal conditions and intelligent parking. The analysis here is based on averages.

The exact analysis was based on a renewal model which was developed in the study and which confirmed the conclusion of the foregoing argument with a high probability.

The next problem involves counting outcomes, and uses a very nice method to accomplish this.

Example 2. The Box-office Cash Problem

There are $2n$ people lined up at a ticket office: n have only \$5 bills; n have only \$10 bills. The box office has no cash when it opens, and each customer will purchase one \$5 ticket. What is the probability that no customer waits for change?

Here, a picture helps considerably. Locate each customer at the points $1, 2, \ldots, 2n$ along the abscissa of an (x, y) coordinate system in the order they appear in line; the box office is at the origin. To each person with a $10 bill, assign $+1$; to each person with a $5 bill, assign -1. As we move from left to right, the values assigned to the customers at the corresponding points are accumulated and plotted on the ordinate. Connecting these ordinates yields a trajectory for the given customer ordering. The trajectory always ends at the point $(2n, 0)$ (Fig. 2.2).

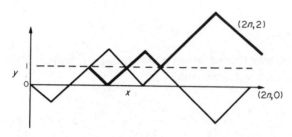

FIG. 2.2.

There are $\binom{2n}{n}$ trajectories (i.e., the number of ways of distributing n ascents among $2n$ ascents and descents). The x-axis represents the condition of no change at the box office, so any ascent above the x-axis represents the arrival of a customer with $10 when there is no change available; all such unfavorable trajectories must touch the line $y = 1$ at least once. To count these unfavorable outcomes, we construct a new trajectory as follows:

(i) The new and old trajectories coincide up to the first point where $y = 1$ is touched.
(ii) From that point on, the new trajectory is a reflection of the old about the line $y = 1$ (see bold line in the figure). The new trajectory must end at $(2n, 2)$ because it will always involve a reflection of the point $(2n, 0)$ about the line $y = 1$. Thus, there are now two more ascents than descents to give a total of $n + 1$ ascents. Hence there are $\binom{2n}{n+1}$ new trajectories and $\binom{2n}{n} - \binom{2n}{n+1}$ favorable (old) trajectories. The desired probability is [Gnedenko]

$$\frac{\binom{2n}{n} - \binom{2n}{n+1}}{\binom{2n}{n}} = \frac{1}{n+1}.$$

In the next example, we show how estimation may be used to determine abstract epistemological limits.

Example 3. Processing Information [Bremermann]

Using ideas from quantum mechanics and information theory, Bremermann has posed the following conjecture: "No data-processing system whether artificial or living can process more than (2×10^{47}) bits per second per gram of its mass."

This bound has been widely discussed in the literature of cybernetics and general systems theory where it has acquired the name "the Bremermann limit."

The proof is simple. Information must be represented by some kind of marker. Assume that energy levels are used as markers. Assume also that the levels must lie in an interval $(0, E_{max})$ and that the energy levels can be measured to accuracy ΔE. At most, one can distinguish

$$n = \frac{E_{max}}{\Delta E} \tag{2.1}$$

energy levels. The information obtained in measuring a random variable which can take on any of n values, each with probability $p_1 \ldots p_n$ is

$$H(p_1 \ldots p_n) = - \sum_{i=1}^{n} p_i \log_2 p_i,$$

which takes on its maximum value when $p_1 = p_2 = \ldots p_n = 1/n$. This maximum is given by H_{max} where

$$H_{max} = \log_2 n. \tag{2.2}$$

We consider how to divide up energy E_{max} to maximize the information. For one marker (per instant) with n levels, there are $n + 1$ distinguishable marker values (counting 0) and, by (2.2),

$$H_{max} = \log_2 (n + 1).$$

Now consider two markers, each with energy levels in the range $(0, \frac{1}{2}E_{max})$. Then a maximum of

$$H_{max} = 2 \log_2 \left(\frac{n}{2} + 1 \right)$$

bits can be represented with energy E_{max}. The optimal use of the given amount of energy E_{max} occurs with n markers with values in $(0, \Delta E)$. Then

$$\max H = n \log_2 \left(\frac{n}{n} + 1 \right) = n \text{ bits.} \tag{2.3}$$

The maximum energy any self-contained computing system of mass m can utilize as a marker is given by Einstein's equation:

$$E_{max} = mc^2 \equiv m \times 9 \times 10^{20} \, cm^2 \, sec^{-2} < m \times 10^{21} \, cm^2 \, sec^{-2}. \tag{2.4}$$

By Heisenberg's uncertainty principle, the accuracy of energy measurement is bounded in time Δt by the relation

$$\Delta E \, \Delta t \geqslant h/2\pi. \tag{2.5}$$

(ΔE = energy uncertainty, Δt = duration of measurement, h = Planck's constant.) Substituting (2.1) and (2.4) into (2.5) yields

$$\frac{E_{max}}{n} \Delta t \geqslant h/2\pi$$

and

$$\frac{mc^2}{(h/2\pi)} \geqslant \frac{n}{\Delta t}$$

which for (2.3) is the optimal energy usage.

Thus, since

$$\frac{mc^2}{(h/2\pi)} \sim 2m \times 10^{47} \text{ bits/sec}$$

we have the desired bound (per gram of mass m) of 2×10^{47} bits/sec.

2.5. Structuring

Possibly the most important factor in modeling is the ability to grasp the essential structure of a problem or situation.

Our first example which is non-numerical refers to a popular puzzle: its solution involves decomposition into simple elements, and the invariance of essential properties under transformation.

Example 1. The Colored Cube Problem [Busacker and Saaty]

An illuminating illustration of a nice way to structure a problem is Instant Insanity. We are given four cubes, the faces of which are colored with one of four colors. Since there are six faces, each cube would have more than one face colored with the same color. The coloring of the cubes is different so two cubes are not necessarily colored the same way. The problem is to stack the cubes vertically in such a way that we have a rectangular prism and each of the rectangles on a side of the prism has all four colors appearing in it, not necessarily in the same order.

Of course, the problem may not have a solution. For example, if all three faces meeting at a corner of each cube are colored with the same color for all four cubes, then that color appears in the stack at least eight times instead of the required four; the problem has no solution.

Our first step is to represent the geometry of the problem by a plane figure. We draw four points as corners of a square letting each corner represent a different color, and we also label the cubes 1 to 4. Each cube has three pairs of opposite faces. We take the first cube and draw lines connecting the vertices associated with each pair of opposite faces and label these three lines with number 1. We do the same for each of the other three cubes. In all, we have 12 lines with three labeled 1, three labeled 2, etc. A possible representation is illustrated in Fig. 2.3 where the colors are labeled R, G, W, and B (for red, green, white and blue).

Next we decompose the problem into two parts: we solve the problem for one pair of opposite reactangles in the stacking and then for the other pair. We note that in each pair each cube is used once and each color appears twice, one in each rectangle. Thus, we draw the four points associated with the colors and choose four lines from the diagram that are labeled 1, 2, 3, 4 (thus each cube is used once) but in such a way that each point is incident with two lines. When this is done, the same procedure is repeated for the remaining pair of opposite rectangles. Adjustment in the choice of lines for the first pair of rectangles may be necessary to solve the problem for the second pair. The solutions for the two parts are independent. Given the solution of one partition, the other is obtained by a rotation around an axis passing through the centers of the two opposite faces of the cubes which appear in the first solution thus leaving it invariant. Figure 2.4(a) and (b) portrays these solutions corresponding to Fig. 2.3. Note that to ensure the alternate appearance of each color in opposite rectangles one can proceed sequentially from one vertex to an adjacent vertex using the cubes in the order in which the edges travel but alternating the distribution of each color between the two opposite rectangles.

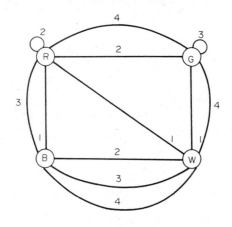

Fig. 2.3.

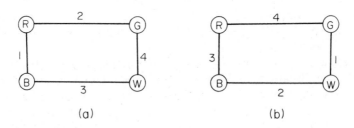

(a)   (b)

Fig. 2.4.

In the next example we use analytical structuring.

Example 2. The Stable Table [Saaty, 1972]

Consider a square table with four legs whose lengths are equal. Suppose that the ground is not a smooth, flat surface, but a wavy, humpy one (not too much relative to the length of the table's legs). Show that there is always a position where the table could stand so that each leg rests on the ground (i.e., the table has no wobble, although it may be tilted).

We note that it is always possible to stand the table so that three legs touch the ground: hold two legs up and tilt the table so that two legs touch the ground. Then put down the other two legs; one of them will touch first. Thus, there is always a pair of legs on one of the two diagonals which touches the ground. Let x be the sum of the heights of these legs above the ground. Since both of them touch the ground, we have $x = 0$. Let y be the sum of the heights of the two legs on the other diagonal (one of them contributes zero height because three legs always touch the ground). Let us rotate the table 90 degrees so that each diagonal goes into the other. Now x and y are functions of the angle of rotation θ, and we have

$$x(0) = 0, \quad y(0) > 0,$$

$$x\left(\frac{\pi}{2}\right) > 0, \quad y\left(\frac{\pi}{2}\right) = 0,$$

$$x(\theta)y(\theta) = 0, \quad 0 \leqslant \theta \leqslant \frac{\pi}{2}.$$

The last relation holds because three legs always touch the ground.

Consider the value of θ at which $x(\theta)$ just changes from zero to positive. At that value $y(\theta)$ must just turn zero. Thus, at that value of θ, both x and y vanish, and all four legs must rest on the ground.

The following example gives a way of structuring a situation so as to create stability. It shows that stability may be achieved by compromise. There are times when our best preference would cause us unhappiness.

Example 3. The Stability of a Set of Marriages [Gale and Shapley]

We consider n boys and n girls to be paired off in marriages: in this social situation we call such a set of marriages unstable if under it there is a boy or a girl who are not married to each other but prefer each other to their actual mates. Otherwise the set of marriages is called stable. The key structural ingredient in this example is the linear ordering by each person of his possible mates. This allows us to find an algorithm for the pairing-off process that produces a solution which is stable by its construction. Each girl (and each boy) ranks the boys (the girls) according to preference, and each boy proposes to his favourite girl. She keeps on a string, but does not yet accept, the man she most prefers. The rejected boys choose again according to their second preferences, etc., and each time the girl keeps on a string the boy she prefers most and releases any less-preferred suitors who may be already on her string. There are at most $n^2 - 2n + 2$ such attempts of proposal by all the boys to all the girls. No boy proposes to the same girl more than once. Hence, every girl gets a proposal. The process terminates when every girl has been chosen (which must obviously be the case). Then the girls accept these boys for marriage. These marriages are stable because if a boy prefers another girl to his wife then she does not prefer him because he would have proposed to her before proposing to his wife but she rejected him in favor of her husband.

We move back in time to the nineteenth century and the buffalo. Our next example involves algebraic structuring of a problem and yields numerical answers.

Example 4. Controlled Slaughter of Animals for Food [Truxal and David]

Between 1830 and 1887 the buffalo population in the United States was reduced from 40 million down to 200 heads. The average weight of a buffalo was about 1000 lbs, an amount of meat sufficient for at least five people for a year. The slaughter was carried out for parts of the animal; its tongue and its hide. In about 60 years, the lack of a policy for harvesting this potential source of food led to an almost complete destruction of the animal.

We know that 10% of the mature animals die each year, maturity is reached in two years, and 90% of the mature females have a calf a year. Of the total calves born, 53% are male and 47% are female. Because of high infant mortality, only 30% of the calves live to maturity.

It is desired to determine the number of beasts to be harvested each year. By controlling the female population, the number of new births is thereby determined. This enables us to estimate the percentage of beasts that should be harvested each year.

If we let F_n denote the number of mature females at the start of year n, then since 10% die and $p\%$ are harvested, $\{0.9 - (p/100)\} F_{n-1}$ survive from year $n-1$ to year n.

In addition, consider the number of mature females F_{n-2} at the beginning of year $n-2$: 90% of them give birth to calves of which 47% are female. Of this group of calves, 30% survive to maturity two years later. Thus these females in year $n-2$ contribute $0.9 \times 0.47 \times 0.3 = 0.1269$ new females to the female population at the beginning of year n. We thus have the following relation:

$$F_n = \left(0.9 - \frac{p}{100}\right) F_{n-1} + 0.1269 F_{n-2}.$$

If it is desired to maintain the same number of mature females every year, $F_n = F_{n-1} = F_{n-2}$ giving $p = 2.69\%$. If this rate had been applied to a population of 40 million males and females (more males can be harvested because there are more of them), more than a million could have been harvested each year, providing meat for at least five million people.

We now turn to a very powerful method for structuring problems, particularly in physics.

*Example 5. Dimensional Analysis [Bridgeman]

In physics, physical variables originate largely in the operations symbolized in their dimensional formulas. Dimensional formulas and equations have a structure closely related to the operations of physical measurement. A small number of variables is selected and all other variables may then be expressed as a function of the basic set.

Usually mass m, length l, time T, and electric charge Q are used as the fundamental, or primary, variables. A theorem in dimensional analysis asserts that any physical variable f is proportional to a product of powers of primary variables $P_1, P_2, \ldots, P_n$; that is, if $a_1, a_2, \ldots, a_n$ are rational numbers, then

$$f = K P_1^{a_1} P_2^{a_2} \ldots P_n^{a_n},$$

where K is a proportionality constant. Frequently K is dropped and one writes $[f] = [P_1^{a_1} \ldots P_n^{a_n}]$ where the brackets indicate dimensional equivalence.

For example, force may be expressed in terms of the primary variables given above as $F = K(MLT^{-2})$.

As an illustration, suppose that it is required to calculate the period of a simple pendulum consisting of a rigid rod of length h supporting a mass m and having an angular displacement θ.

The functional expression for the physical variables involved is

$$t = t(m, h, g, \theta),$$

or, dimensionally,

$$[T] = [M^a L^b (LT^{-2})^c].$$

The equations obtained by equating corresponding powers are

$$M:0 = a,$$
$$L:0 = b + c,$$
$$T:1 = -2c.$$

Therefore $a = 0, \quad b = \frac{1}{2}, \quad c = -\frac{1}{2}.$

Therefore $t = k(h^{1/2}/g^{1/2}).$

Since we wish to take θ into account, we include an unknown function $f(\theta)$ and write

$$t = K \sqrt{\frac{h}{g}} f(\theta).$$

Experimentally, one finds that, if θ is small, $f(\theta)$ is almost constant and approximately 2π (assuming $K = 1$). Note that since the exponent of mass is zero the period is independent of the mass.

Thus $t = 2\pi \sqrt{h/g}.$

By similar arguments the reader will have no difficulty in verifying the well-known elementary relations of physics, $s = \frac{1}{2}gt^2$ and $v^2 = 2as$, where s is distance, v is velocity, a is acceleration, and g is the acceleration due to gravity.

Dimensional analysis plays an important role in forming "order of magnitude" estimates of certain properties of physical systems. We give two illustrations.

(i) The Power in Wind

Suppose the wind blows with velocity V at a windmill whose area facing the direction of the wind is A. The only relevant physical constant we need is ρ, the density of air. The dimensions are:

$$[V] = LT^{-1},$$
$$[A] = L^2,$$
$$[\rho] = ML^{-3}.$$

The dimension of power P (energy $\times T^{-1}$) is

$$[P] = [ML^2 T^{-3}] = [V]^a [A]^b [\rho]^c$$

from which $c = 1, \quad a = 3, \quad b = 1,$

and $[P] = [V^3 A\rho].$

(ii) Ship Size and Speed

A powerful illustration of the use of this method now follows.

Let a ship make a voyage, e.g., across the Atlantic at a speed V. Let L be the length of the ship, S its wet surface area, T its tonnage, D its displacement or volume, H its horsepower, R the resistance, and C the coal necessary for the voyage. It is known from experiments comparing two ships that $L \propto V^2$. If we double the length of a ship from $L_1 = 1$ to $L_2 = 2$, then the corresponding speeds satisfy $V_1^2/V_2^2 = 1/2$. Now we have

$$
\begin{aligned}
[V] && [V] \\
[L] && [V^2] \\
[S] &= [L^2] & [V^4] \\
[D] &= [T] = [L^3] = [V^6] \\
[R] &= [SV^2] &= [V^6] \\
[H] &= [RV] &= [V^7] \\
[C] &= [H/V] &= [V^6]
\end{aligned}
$$

Thus, if it is desired to increase the speed by 1 %, the length must be increased by 2 %, the displacement by 6 %, the coal consumption by 6 %, the horsepower and the necessary boiler capacity by 7 %.

2.6. Proving

Many problems may be solved by proving or disproving a given hypothesis; this is one of the most common processes in mathematics.

In the legal field one is often concerned about being fair in judgment or division of assets. Here is a simple creative illustration of how to do fair division.

Example 1. Fair Division [Saaty, 1970]

How might one fairly divide a cake among three people so that each is satisfied with his share?

If there were two people only, one of them divides it and the other chooses his share first.

For three people, we let the first one divide the cake into three parts and ask the second to choose a piece which he keeps if the third person does not object on the ground that it is too large. If he objects, he is asked to cut this part into two parts of which the second chooses one and the third person takes the other. Of the two remaining parts, either both the second and the third agree on one piece which they divide or they each want to divide a different part with the first person. In either case, the problem is reduced to dividing a cake fairly between two people.

A common method of proof is that of mathematical induction. This requires that one must show that a given property holds for $n = 1$ and that whenever the property is assumed true for $n = k$ it is also true for $n = k + 1$. The principle of induction then says that the property is true for all n. Prove by induction that $1 + 2 + \ldots + n = \dfrac{n(n+1)}{2}$.

Example 2. Acquaintances

Given six people, each of whom may or may not know any of the others, prove that there is always either a set of three people who know each other or three people none of whom knows the other two.

Suppose this statement is false; i.e., there are never three people who either do or do not know one another. Consider now any of the six and that person's relations with the other five. These relations involve knowing or not knowing each of them. Because there are five there will always be either at least three people whom he knows or three people whom he does not know. Suppose he knows at least three people and consider any three of them. Since the statement is assumed false, each of them must not know the other two, for otherwise, we would have a triangle of two of them and the original one with which we started, who know each other, which would contradict the assumption that the statement is false. Now, this group of three people who do not know each other violates the assumption that the statement is false. Thus, the assumption is itself false, and the statement must be true. A similar argument would apply if he did not know at least three people.

Note that the statement would not be true if we started out with five people: they might be represented as situated at the nodes of a pentagon, each connected to its two neighbors for acquaintance but to no other. It would not be possible in this situation to find the triangle in which three people know one another or do not know one another.

This result can also be placed in the context of graph theory. Let K_n denote the graph consisting of n points and all the possible edges joining distinct pairs of points. No matter how we color the edges of K_6 using two colors, say red and blue, we find that there must be a triangle of edges all colored the same (such a triangle is called *monochromatic*). (We can let the six points correspond to the six people with red lines to denote acquaintanceship and blue lines nonacquaintanceship.) It is suggested at this point that the reader draw a number of possible figures as illustrations of the situation. We can show what can happen for five people by drawing a blue five-pointed star and circumscribing a red pentagon. This drawing of K_5 has no monochromatic triangle.

This completes our introduction to the very fundamental topics of counting, estimation, and structuring. The references at the end of this chapter contain a wide variety of examples.

In our next chapter, we shall consider different types of equations and the ways in which they arise in modeling.

Chapter 2—Problems

1. What is the maximum number of pieces into which a circular pie can be cut with n straight cuts if no piece can be moved after any cut? Answer: $(n^2 + n + 2)/2$. (*Hint:* relate the number of pieces resulting from n cuts to those resulting from $n - 1$ cuts.)

2. Two missionaries and two cannibals wish to cross from the left to the right bank of a river by row boat. The boat can carry no more than two people at a time and must be brought back for the next crossing. At no time should there be more cannibals than missionaries, since the cannibals, from force of habit, would eat the outnumbered missionaries. How can the crossing take place? (*Hint:* represent the states on the left bank by (a, b) where a is the number of missionaries b is the number of cannibals, and consider means of transition from $(2, 2)$ to $(0, 0)$. What transitions are possible? Can you draw a graph? Can you represent this graph by a matrix? Would it be simpler to solve this problem by trial and error? What would happen if the numbers were larger?)

3. We want to find the most efficient method of parking cars in a lot. How can we structure this problem?

(Consider the simpler problem of packing circles in a plane so that any circle may be moved by sliding without disturbing any of the other circles.) For a full discussion of this problem, see Heppes.

4. Take the marriage stability example and translate it to the problem of multiple applications by n students to m colleges with $m \leqslant n$. (Note that a college may accept more than one student.)

5. Telephoned orders are placed alphabetically according to the family name. There is thus a bias towards letters like S and when the order arrives there is more search to do under S. A new system uses the last two numbers of the customer's phone number. These are assumed to be random and hence would distribute the customers' orders uniformly in the catalogue and could be accessed more rapidly. Can you propose two other ways for this purpose? What are other activities for which you have more systematic and efficient implementations?

6. Most theatres empty their audience in a matter of 10–15 min. You are standing at the door of a theatre observing the rate at which people exit. The theatre has four exit doors. Give a reasonable way to estimate the number of seats without going in.

7. How would you estimate the average number of hairs on a person's head?

8. Give a formula for estimating the number of uniform spherical cherries of a certain radius to be packed in a cylindrical can of given radius and height. How would you allow for space between cherries? Enter this in your model. Do this a second time with a different scheme.

9. Give an efficient, approximate hierarchical scheme for getting 10 people to count a stack of money consisting of about $1,000,000. (If all else fails, weight it.)

10. Use dimensional analysis to obtain an expression for the power which could be extracted per unit length of coastline facing the incoming sea. [*Hint*: The relevant dimensions include wavelength, period, the density of water, the amplitude and gravity.]

11. Given the number of digits D used in printing the page numbers of a text, determine the number of pages P in the text. For example, a text of 13 pages uses nine digits up to page 9 and two digits per page from pages 10 to 13 yielding a total of 17 digits. The problem here is to reverse this process. Show that

$$P = \left(D + \frac{10^{m+1} - 1}{9} \right) / (m+1) - 1.$$

12. The classical counting problem known as the hat-check problem may be formulated as follows: n men walk into a bar, check their hats, and proceed to get drunk. They all leave together but, fumbling with their hat-checks, wind up drawing hats at random. Show that the probability that no man gets his own hat is

$$\frac{1}{2!} - \frac{1}{3!} + \frac{1}{4!} - \cdots \frac{(-1)^n}{n!}.$$

(*Hint*: let A_n be the number of ways of distributing n hats among n men such that no one gets his own hat and P_n be the corresponding probability. If $(n-1)$ men go through the process and end with no man having his own hat and if an nth man joins them with his hat then he can exchange it with any one of them. This gives $(n-1)A_{n-1}$ ways. Further, if any one of them has his own hat, he can exchange it with the new man. This adds $(n-1)A_{n-2}$ ways. There are no other possibilities. Thus $A_n = (n-1)(A_{n-1} + A_{n-2})$. Also $A_i = i! \, p_i$.)

13. Two diagonally opposite corner squares are cut out from a chess board. Is it possible to cover the remainder of the board with nonoverlapping dominos, each of which covers two squares? [*Hint*: Consider the color of the squares which are cut out.] Straddling the left most column and its neighboring column, there must be an odd number of dominos. Since the neighbor gets an odd number straddling it and its right neighbor there is also an odd number and so on. In all the number of horizontal dominos is odd. Similarly the number of vertical dominos is odd yielding an even total number. However, 31 is all that we need.

References

Asimov, Isaac, *The Universe: From Flat Earth to Quasar*, Walker, New York, 1966.

Bridgeman, P., *Dimensional Analysis*, Yale University Press, 1931.

Bremermann, H. J. Optimization through evolution and recombination, in *Self-Organizing Systems* (eds. M. C. Yovits, G. T. Jacobi, and G. D. Goldstein), Spartan Books (Washington DC, 1962), p. 93.

Busacker, Robert G. and Thomas L. Saaty, *Finite Graphs and Networks*, McGraw-Hill, New York, 1965.

Few, L., The shortest path and the shortest road through n points, *Mathematika*, Vol. II, 1955 pp. 141–144.

Gale, D. and L. S. Shapley, College admissions and the stability of marriage, *Am. Math. Monthly*, Vol. 69, No. 1, 1962.

Gnedenko, B. V., *The Theory of Probability*, Chelsea Publishing Co., New York, 1951.

Heppes, A., On the densest packing of circles not blocking each other, *Stud. Sci. Math. Hung.*, Vol. 2, No. 1–2, 1967 pp. 257–263.

Saaty, Thomas L., *Optimization in Integers and Related External Problems*, McGraw-Hill, New York, 1970.
Saaty, Thomas L., *The Thinking Man's Joke Book Series*, Vol. IV, G. & S. Publications, 1972.
Truxal, J. G. and E. E. David, Jr., *Man and His Technology*, Engineering Concepts Curriculum Project, McGraw-Hill, New York, 1971, 1973.

Bibliography

Berge, Claude, *The Theory of Graphs and Its Applications*, Wiley, New York, 1962.
Berge, Claude, *Principles of Combinatorics*, Academic Press, New York, 1971.
Cohen, M. R. and E. Nagel, *An Introduction to Logic and Scientific Method*, Harcourt, Bruce & Co., New York, 1934.
Courant, Richard and Herbert Robbins, *What is Mathematics?*, Oxford University Press, 1941.
Feller, W., *An Introduction of Probability Theory and Its Application*, Vol. I, 2nd edn., Wiley, New York, 1957.
Landau, E., *Foundations of Analysis*, Chelsea Publishing Co., New York, 1951.
Liu, C. L., *Introduction to Combinatorial Mathematics*, McGraw-Hill, New York, 1968.
McMahon, Percy A., *Combinatory Analysis*, Vols. 1 and 2, Chelsea Publishing Co., New York, 1960.
Netto, E., *Lehrbuch der Combinatorik*, Chelsea Publishing Co., New York, 1901.
Parzen, Emanuel, *Modern Probability Theory and Its Applications*, Wiley, New York.
Riordan, J., *An Introduction to Combinatorial Analysis*, Wiley, New York, 1958.
Ryser, Herbert J., *Combinatorial Mathematics*, Wiley, New York, 1963.
Saaty, Thomas L., *Mathematical Models of Arms Control and Disarmament*, Wiley, New York, 1968.
Saaty, Thomas L., (ed. with F. J. Weyl), *The Spirit and Uses of the Mathematical Sciences*, McGraw-Hill, New York, 1969.

PART II

FRAMEWORK

EQUATIONS play a dominant role in structuring problems. An equation is a statement of identity or a conditional equality (holding only for certain values of the unknowns) between two quantities. Equations usually involve two types of quantities: parameters and variables. The parameters are known or controlled numerical values which enter into the formulation. The variables are unknown and are allowed to assume any value in a specified range. The problem is to determine these unknowns. A system of equations may be underdetermined if it has more variables than the number of equations and over-determined in the opposite case. In the former case, some of the variables are arbitrary and the remaining variables may be expressed as functions of the arbitrary variables. Ordinarily, if the system of equations is not redundant and the number of variables is equal to the number of equations, it is possible to obtain specific values for the variables as solutions. The values of the variables depend on those of the coefficients, and solutions of equations are thus functions of their parameters.

An optimization problem has a structure involving a set of constraints given in the form of equations or inequalities and a function, frequently called an objective function to be maximized or minimized subject to the constraints. Optimization is a normative or prescriptive field in which among all possible solutions which satisfy the system of constraints, that which yields a maximum or minimum (called an optimum) to the objective function is sought. Often, the process of solving an optimization problem can be reduced to a process of solving a set of equations.

Stochastic processes can be visualized formally in terms of probability theory and equations. Even when the structure of a problem is probabilistic, we use equations and inequalities for its representation. In this case the variables of the problem are subject to the laws of probability, and the method of solution differs from that of a nonprobabilistic problem; the parameters of the equations may also be subject to randomness, adding to the complexity of the solution. Optimization problems in which probability plays a central role have come to the fore in recent years; such formulations are thought to be closer to reality and to give more faithful representations than non-probabilistic equations. So, the three pillars—equations, optimization, and probability—can be seen to be closely interdependent in the process of modeling.

Chapter 3

Equations

3.1. Introduction

OUR purpose here is to give some applications for each of the five basic types of equations (and of their hybrids). These types are: algebraic, differential (ordinary and partial), difference, integral, and functional. Diophantine equations, algebraic equations with rational coefficients whose solutions must be integers, are also illustrated. Each of the above types of equations may also have stochastic elements; this type of problem often arises in modeling. We do not illustrate these extensively in this chapter. Hybrid equations include differential–difference, integro-differential, integro-difference, integro-differential–difference, etc.

There are no clear dividing lines between the examples in this chapter and those in others, especially since equations and inequalities are so central to all quantitative mathematics. However, the purpose of this chapter is to catalogue important types of equations. Few of the equations are solved; we are concerned essentially with their formulation. A comprehensive treatment of most types of equations is given in *Nonlinear Mathematics* by Saaty and Bram and *Modern Nonlinear Equations* by Saaty included in the bibliography (now to appear in paperback by Dover Publications).

3.2. Algebraic Equations

Very briefly an equation of the form

$$a_0 x^n + a_1 x^{n-1} + \ldots + a_n = 0 \quad (a_0 \neq 0),$$

where $a_0, a_1, a_2, \ldots, a_n$ are real or complex numbers, is called an algebraic equation of degree n. We know from the fundamental theorem of algebra (due to Gauss) that for every algebraic equation of degree n there exists n *roots* or *solutions*.

Frequently we are interested not in a single equation but in a system of equations. Among the simplest systems of algebraic equations is a linear system of m equations and n unknowns $x_1, x_2, \ldots, x_n$ given by

$$a_{11}x_1 + a_{12}x_2 + \ldots + a_{1n}x_n = b_1,$$
$$a_{21}x_1 + a_{22}x_2 + \ldots + a_{2n}x_n = b_2,$$
$$\ldots \ldots \ldots \ldots \ldots \ldots \ldots \ldots \ldots \ldots,$$
$$a_{m1}x_1 + a_{m2}x_2 + \ldots + a_{mn}x_n = b_m.$$

In algebra the existence of a solution of such a system is given by means of the concept of the rank of the matrix of coefficients and the rank of the augmented matrix which includes the right side constants as an additional column. This matrix is not square. The details would take us far afield, but this system is *solvable* if and only if the rank of the matrix of the system is equal to the rank of the augmented matrix. A system of linear equations is said to be *overdetermined* if the number of variables n is less than the number of linearly independent equations m. It is *determinate* if $m = n$ and underdetermined if $m < n$.

If $b_i = 0$ $(i = 1 \dots m)$, then we have a system of m homogeneous equations in n unknowns. If $m = n$, this system has a nontrivial solution if and only if the determinant of the matrix of the system is equal to zero.

Systems of nonlinear algebraic equations are usually solved by *numerical approximation* methods such as Newton's method and gradient methods.

We give a simple example which involves algebraic equations.

Example. Strategy for a Mailman

A mailman must deliver mail to each house on both sides of a straight street of length L and width W. Suppose that all N houses have the same street frontage of length D. The houses on both sides may be considered as points in the plane and start precisely at the beginning of the street and end at its end. The houses on one side are a mirror image of those on the other. If the mailman crosses the street he does so on a perpendicular to its length. Compare the strategy of delivering mail to all houses on one side, crossing the street and returning to his starting point by delivering to all houses on the other side, with the strategy of crossing the street from one house to that opposite it on the other side, walking to its next door neighbor and crossing the street again.

Note that
$$\frac{N}{2} - 1 = \frac{L}{D}.$$

We want conditions on D and W such that the mailman would follow the first or second strategy to walk the least distance. Using the first strategy he walks $2L + W$ and, using the second, he walks $(N/2) W + L$.

The problem is: when is $2L + W < L + (N/2) W$ which is equivalent to $L < [(N/2) - 1] W$. Substituting $L = [(N/2) - 1] D$ gives $[(N/2) - 1] D < [(N/2) - 1] W$ or $D < W$. Thus, he uses the first strategy when $D < W$ and the second when $D \geqslant W$.

3.3. Diophantine Algebraic Equations

Sometimes the solutions of an algebraic equation must be integers. A simple example follows.

Example. Economic Constituency of an Air Fleet

An airline buys Boeing 707 aircraft at $\$12 \times 10^6$ each, Boeing 727s at $\$8 \times 10^6$ each, and small executive type jets at $\$2 \times 10^6$ each. If a total sum of $\$8 \times 10^7$ is paid for 20 aircraft, including at least one of each kind, how many aircraft of each kind does the airline buy?

If x, y, z are the respective numbers bought of each kind of aircraft mentioned in the problem, then one must find positive integers x, y, z which satisfy $x + y + z = 20$, $6x + 4y + z = 40$. Subtraction gives $5x + 3y = 20$. For this equation to hold, y must be divisible by 5. Thus we begin by choosing $y = 5$ from which we get $x = 1$ and $z = 14$. In fact this is the only possible solution.

3.4. Ordinary Differential Equations

Recall that a differential equation is an equation involving functions of one or more *independent* variables and certain of their derivatives. Such an equation is called ordinary if there is one independent variable. The order of a differential equation is the order of the highest order derivative it contains.

A functional relation between the dependent and independent variables which satisfies the differential equation is said to be a solution of the equation. A differential equation is completely solved when all of its solutions are known.

If m differential equations are given in n unknown functions, we speak of a system of differential equations.

By a solution of the system of equations of the first order

$$y'_1 = f_1(x, y_1, y_2, \ldots, y_n),$$
$$y'_2 = f_2(x, y_1, y_2, \ldots, y_n),$$
$$\vdots$$
$$y'_n = f_n(x, y_1, y_2, \ldots, y_n),$$

where

$$y'_i = \frac{dy_i}{dx}, \quad i = 1, 2, \ldots, n.$$

we mean a system of functions

$$y_1 = g_1(x), \ y_2 = g_2(x), \ldots, \ y_n = g_n(x),$$

which satisfies the system simultaneously.

Given this system and a point $P(a, b_1, \ldots, b_n)$, if the functions $f_1, f_2, \ldots, f_n$ are continuous in a neighborhood $N(E)$ of P and have continuous partial derivatives there with respect to the variables $y_1, y_2, \ldots, y_n$, then in a certain neighborhood of a there exists precisely one set of functions which is a solution of the system satisfying the *initial conditions*

$$g_1(a) = b_1, \ g_2(a) = b_2, \ldots, \ g_n(a) = b_n.$$

Sometimes *boundary conditions* are imposed on the solution leading to a boundary-value problem. In general, the number of equations need not be equal to the number of unknowns. Under such general circumstances, the conditions given above on the *existence* and *uniqueness* of the solution need no longer be valid.

We now give some illustrations of models in which ordinary differential equations are used.

Example 1. Advertising

When two competitive companies engage in advertising campaigns, it is reasonable to assume that the rate at which each company increases its sales is proportional to the available market. (The constants of proportionality will depend on the effectiveness of each campaign.) Thus we have the following differential equations:

$$\frac{dS_1}{dt} = C_1 M_a, \quad \frac{dS_2}{dt} = C_2 M_a,$$

where $S_i(t)$ is the sales of company i in any year (S/year), $M_a(t)$ is the available market $= M(t) - S_1(t) - S_2(t)$, $M(t)$ is the total market, and C_1 and C_2 are constants.

The solutions for $S_1(t)$ and $S_2(t)$ will depend on the form for $M(t)$ which on occasion has been taken to be

$$M(t) = \alpha(1 - e^{-\beta t}),$$

where α and β are constants.
We have,

$$\frac{dS_2}{dt} = C_3 \frac{dS_1}{dt},$$

$$S_2(t) = C_3 S_1(t) + C_4,$$

$$M_a(t) = \alpha(1 - e^{-\beta t}) - (1 + C_3)S_1 - C_4,$$

$$\frac{dS_1}{dt} + AS_1 = Be^{-\beta t} + C,$$

where $\qquad A = C_1(1 + C_3), \quad B = -C_1 \alpha e^{-\beta t}, \quad C = C_1 \alpha - C_4.$

Now the first-order linear differential equation

$$\frac{dy}{dx} + p(x)y = q(x)$$

has the general solution

$$y(x) = e^{-\int p(x)dx}\left[K_1 + \int e^{\int p(x)dx}q(x)dx\right].$$

Thus the general form of the solution for $S_1(t)$ is given by

$$S_1(t) = K_1 e^{-At} + K_2 e^{-\beta t} + K_3.$$

A similar expression may be obtained by the reader for $S_2(t)$.

Example 2. Lanchester's War Model

The following differential equation model, developed by Lanchester, is concerned with the problem: Suppose that N_1 units of one force A, each of hitting power α, are engaged with N_2 units of an enemy B, each of hitting power β. Suppose further that the engagement is such that the fire power of force A is directed equally against all units of B and vice versa. The rate of loss of the two forces is given by

$$\frac{dN_1}{dt} = -k_2 \beta N_2 \quad \text{and} \quad \frac{dN_2}{dt} = -k_1 \alpha N_1,$$

where k_1, k_2 are positive constants. (N_1 and N_2 are, of course, integers, but it is convenient to treat them here as continuous variables.) The strength of the two forces is defined as equal when their fractional losses are equal, that is, when

$$\frac{1}{N_1}\frac{dN_1}{dt} = \frac{1}{N_2}\frac{dN_2}{dt}.$$

This type of analysis is used in balance of power problems. On dividing the first equation by the second and integrating, we obtain two hyperbolas defined by $\alpha N_1^2 - \beta N_2^2 = C$. Taking $C = 0$ gives Lanchester's N^2 law, which states that the strength of a force is proportional to the fire power of a unit multiplied by the square of the number of units.

3.5. Partial Differential Equations

An equation of the form

$$F\left(x_1, x_2, \ldots, x_n, z, \frac{\partial z}{\partial x_1}, \ldots, \frac{\partial z}{\partial x_n}, \frac{\partial^2 z}{\partial x_1^2}, \ldots, \frac{\partial^2 z}{\partial x_n^2}, \ldots\right) = c,$$

which relates the unknown function $z(x_1, x_2, \ldots, x_n), (n \geqslant 2)$ and its derivatives is called a partial differential equation. The order of the highest order derivative appearing in the equation is called the order of the equation. In general, we may consider a system of partial differential equations for the unknown functions $z_1, z_2, \ldots, z_r$. If the number of equations is greater than the number of functions to be found, then, in general, there is no system of functions to satisfy all the equations. In this case, the equations are said to be *incompatible*. Certain conditions, the so-called conditions of integrability, must be satisfied in order for the solution to exist. For example, a necessary condition for the system

$$\frac{\partial z}{\partial x} = A(x, y, z),$$

$$\frac{\partial z}{\partial y} = B(x, y, z),$$

with $z \equiv z(x, y)$ to be solvable is that the mixed second partial derivatives $\partial^2 z/\partial y\,\partial x$ and $\partial^2 z/\partial x\,\partial y$ should be equal. Now

$$\frac{\partial^2 z}{\partial y\,\partial x} = \frac{\partial}{\partial y}\left(\frac{\partial z}{\partial x}\right) = \frac{\partial}{\partial y}(A) = \frac{\partial A}{\partial y} + \frac{\partial A}{\partial z}\frac{\partial z}{\partial y},$$

$$\frac{\partial^2 z}{\partial x\,\partial y} = \frac{\partial}{\partial x}\left(\frac{\partial z}{\partial y}\right) = \frac{\partial}{\partial x}(B) = \frac{\partial B}{\partial x} + \frac{\partial B}{\partial z}\frac{\partial z}{\partial x}.$$

If we impose equality and substitute

$$\frac{\partial z}{\partial y} = B \quad \text{and} \quad \frac{\partial z}{\partial x} = A$$

we obtain the necessary condition

$$\frac{\partial A}{\partial y} + \frac{\partial A}{\partial z}B = \frac{\partial B}{\partial x} + \frac{\partial B}{\partial z}A.$$

(A, B, $\partial A/\partial y$, $\partial A/\partial z$, $\partial B/\partial x$, $\partial B/\partial z$ are assumed to be continuous.)

We give one example here. Other examples can be found in later chapters.

Example. Spending Money for the Taxi Driver's Wife

Every time a taxi driver earns a total sum of μ dollars, he gives it to his wife. The time for earning this sum is a random variable, exponentially distributed with mean $1/\lambda$. His wife spends a dollar per day, but only as long as she has money. She does not accumulate any debts. Let ($F(x, t) = $ prob [the wife has an amount of x or less dollars at time t]; determine $F(x, t)$.

Clearly, $F(x, t) = 0$ for $x < 0$.

Since the time to earn μ dollars is exponentially distributed,

$$F(x, t+h) = (1 - \lambda h) F(x+h, t) \quad \text{for} \quad 0 \leqslant x < \mu - h, \tag{3.1}$$

$$F(x, t+h) = (1 - \lambda h) F(x+h, t) + \lambda h F(x+h-\mu, t) \quad \text{for} \quad x \geqslant \mu - h, \tag{3.2}$$

By Taylor's expansion,

$$F(x+h, t) = F(x, t) + h \frac{\partial F(x, t)}{\partial x} + o(h).$$

Substituting in (3.1) and (3.2) yields

$$\frac{F(x, t+h) - F(x, t)}{h} = -\lambda F(x, t) + (1 - \lambda h) \frac{\partial F(x, t)}{\partial x} + \frac{o(h)}{h} \quad (0 \leqslant x < \mu - h), \tag{3.3}$$

$$\frac{F(x, t+h) - F(x, t)}{h} = -\lambda F(x, t) + (1 - \lambda h) \frac{\partial F(x, t)}{\partial x} + \lambda F(x+h-\mu, t)$$

$$+ \frac{o(h)}{h} \quad (x \geqslant \mu - h). \tag{3.4}$$

By letting $h \to 0$ we obtain the equations

$$\frac{\partial F(x, t)}{\partial t} = -\lambda F(x, t) + \frac{\partial F(x, t)}{\partial x} \quad (0 \leqslant x < \mu), \tag{3.5}$$

$$\frac{\partial F(x, t)}{\partial t} = -\lambda F(x, t) + \frac{\partial F(x, t)}{\partial x} + \lambda F(x-\mu, t) \quad (x \geqslant \mu), \tag{3.6}$$

which may then be solved to find $F(x, t)$.

3.6.   Difference Equations

An ordinary difference equation of the kth order has the form

$$x_{n+k} = f(n, x_n, x_{n+1}, \ldots, x_{n+k-1}),$$

where x is a real variable which depends on an integer valued variable n and k is an integral increment added to n. As with differential equations, the solution of a difference equation requires boundary conditions to ensure uniqueness.

The above form may be extended to the case where n is replaced by a real continuous variable t but k remains an integer. Thus, we write $x(t+k) = f[t, x(t), x(t+1), \ldots, x(t+k-1)]$. A system of simultaneous difference equations may take the form

$$f_i(n, x_n, x_{n+1}, \ldots, x_{n+k}, y_n, y_{n+1}, y_{n+2}) = 0 \quad (i = 1, 2).$$

Such a system may sometimes be reduced by elimination to a single equation. The number of equations can be increased by replacing n by $n+1$, for example, and repeating this procedure in order to obtain a sufficiently expanded system to enable elimination of all the y's. Problems which seem to resist easy formulation as algebraic equations may often yield nicely to difference equation methods. We give a simple example. It is taken from real life.

Example. The Fish Aquarium

A fish aquarium contains n units of water. Each week one unit evaporates and must be replaced with fresh water. Because fresh water contains uniformly a certain amount of salt, there is a possibility that the concentration of salt in the aquarium will become dangerously high for the fish. To cut down this risk, at the end of the week, when $n-1$ units are present, another unit is removed from the aquarium (leaving $n-2$ behind) and then two units of fresh water are added. This will still lead to an increase in the concentration of salt, but not as much as if the additional unit were not removed. Prove that ultimately the concentration of salt per unit in the aquarium approaches twice that in fresh water.

Let c be the concentration of salt per unit of fresh water and let X_k be the total amount of salt in the aquarium at the end of the kth week with the level of water brought to normal. We have

$$X_k = X_{k-1} - \frac{X_{k-1}}{n-1} + 2c,$$

$$= \frac{n-2}{n-1} X_{k-1} + 2c,$$

with
$$X_0 = nc,$$

from which it may be seen that the concentration approaches $2c$.

3.7. Differential Difference Equations

We now consider equations in which there are both derivatives and differences; we illustrate with a simple example.

Example. Cascading Cups [*American Mathematical Monthly*]

A series of cups of equal capacity have been filled with water and arranged one below another. Pour into the first cup a quantity of wine equal to the capacity of the cup at a constant rate and let the overflow in each cup go into the cup just below. Assuming that complete mixing of wine and water takes place instantaneously, find the amount of wine in each cup at any time t and at the end of the process at time T.

Let q denote the capacity of each cup, so that the rate of flow is q/T.

Let x_k be the amount of wine in the kth cup and x_{k-1} that in the $(k-1)$st cup. Then the rate of flow of wine into the kth cup is x_{k-1}/T and out of it is x_k/T. This gives

$$\frac{dx_k}{dt} = \frac{x_{k-1}}{T} - \frac{x_k}{T}.$$

Any equation in the sequence can be solved if the one before it has been solved. Thus

$$\frac{dx_1}{dt} = \frac{q}{T} - \frac{x_1}{T},$$

which gives

$$x_1 = q\left(1 - \frac{1}{e^{t/T}}\right),$$

$$\frac{dx_2}{dt} = \frac{q(1 - 1/e^{t/T})}{T} - \frac{x_2}{T},$$

which gives

$$x_2 = q\left[1 - \frac{1 + (1/1!)\,(t/T)}{e^{t/T}}\right],$$

and so on.

For the final amounts of wine in the successive cups one has

$$x_k = q\left(1 - \frac{e_k}{e}\right),$$

where e_k is the sum of the first k terms of

$$1 + \frac{1}{1!} + \frac{1}{2!} + \frac{1}{3!} + \cdots.$$

(Note that as $k \to \infty$, $x_k \to 0$.)

It is not essential that the rate of flow be constant. If x, the amount of wine poured into the first cup, is the independent variable, then

$$\frac{dx_1}{dx} + \frac{x_1}{q} = 1 \ldots \frac{dx_k}{dx} + \frac{x_k}{q} = \frac{x_{k-1}}{q}$$

and

$$x_k = q\left\{1 - \left[1 + \frac{x}{q} + \frac{1}{2!}\left(\frac{x}{q}\right)^2 + \cdots + \frac{1}{(k-1)!}\left(\frac{x}{q}\right)^{k-1}\right]e^{-x/q}\right\}.$$

The final result (at time T) is obtained by setting $x = q$.

3.8.   Delay–Differential Equations

Differential equations with deviating arguments in which the arguments of the highest order derivative of the unknown function are identical and assume values that are never less (more) than those of the unknown function and its remaining derivatives are called equations of *delay*, *retarded*, or *lagging* (advanced or leading) type. We illustrate with an example.

Example. Population Growth—Delayed Harvest

Consider the delay–differential equation

$$\frac{z'(u)}{z(u)} = \alpha - \beta z(u - r) \quad (\alpha > 0).$$

The left side has a form identical to that of the strength of a force in Lanchester's equation. This equation can be used to describe the net birthrate of a population $z(u)$ at time u, and expresses the condition that this birthrate is a constant minus a quantity proportional to the population at a previous time $u - r$.

If we put $u = rt$, multiply both sides of the equation by $r^2\beta$, and define $x(t) \equiv \beta r z(rt)$, the equation now becomes

$$x'(t) = [a - x(t - 1)]x(t),$$

where $a = \alpha r$, which is a slightly simpler form of a delay–differential equation.

3.9.   Differential–Difference–Delay Equations

We now give an example where the equation which is obtained involves derivatives, differentials, and delay in the arguments.

Example Car Following [L. C. Edie.]

Here we describe the simplest situation of identical car following, such as might occur on long stretches of highway in dense traffic when no passing is possible. An approximate description of the way in which a car follows a leader is given by the relative velocity control in which acceleration at time t of the following car is proportional to the relative velocity of the two cars at a retarded time $t - V$ and inversely proportional to the distance between them. This model has been tested in practice and found to be satisfactory.

We may express these conditions as follows:

$$M\ddot{x}_{n+1}(t) = \frac{\lambda[\dot{x}_n(t - \Delta) - \dot{x}_{n+1}(t - \Delta)]}{[x_n(t - \Delta) - x_{n+1}(t - \Delta)]},$$

where M is the mass of car, $x_n(t)$ is the position of car n at time t, Δ is the lag time of system, and λ is the driver–car sensitivity coefficient (constant).

We use the convention that $\dot{x}_n(t)$, $\ddot{x}_n(t)$ denote that one should differentiate once and twice with respect to time and hence denote velocity and acceleration respectively.

We note that this equation may be interpreted both as a dynamic equation of motion and as a stimulus–response equation. If we integrate the equation, we obtain

$$M\dot{x}_{n+1}(t) = \lambda_1 \log\left[\frac{x_n(t - \Delta) - x_{n+1}(t - \Delta)}{L}\right],$$

where L is the length of each car.

3.10. Integral Equations

There are also models in which the relation to be satisfied involves an integral in an unknown function (which has to be determined).

Example. Renewals [Saaty, 1961]

We consider a system with many components. We know from experimental results that the probability of failure of an important component is $f(x)\,dx$ during an interval of time dx. There are N of these particular components in the system, and we need to know the expected number of renewals which will be needed by time T.

We define renewals of the first kind to be renewals of the initial components, renewals of the second kind to be renewals of renewals of the first kind, etc. Then if $u_i(t)$ is the expected number of renewals of the ith kind at time t, we have

$$u_1(t) = Nf(t),$$

$$u_{i+1}(t) = \int_0^t u_i(t-x)f(x)dx,$$

$$i = 1, 2. \ldots$$

We define

$$u(t) = \sum_{i=1}^{\infty} u_i(t).$$

Then the expected total number of renewals $U(t)$ by time T is given by

$$U(T) = NF(T) + \int_0^T dt \int_0^t u(t-x)f(x)dx$$

$$= NF(T) + \int_0^T U(T-x)dF(x),$$

where

$$F(T) = \int_0^T f(x)\,dx.$$

3.11. Integro–Difference and Integro–Differential Equations

We now consider some examples in which the equations which we obtain contain both integrals and differences or derivatives.

Example 1. A Queue

Suppose that customers arrive before a single server and queue up to be served on a first-come, first-served basis. This is known as ordered service. The waiting time (when there is a line) of the $(n+1)$st customer w_{n+1} equals the waiting time of the nth customer w_n plus his service time s_n, minus the time between the arrival of the nth and $(n+1)$st customer which we denote by t_n.

Thus

$$w_{n+1} = w_n + s_n - t_n.$$

Since each of s_n and t_n is given by a probability distribution so are w_n and w_{n+1}. We wish to determine these unknown distributions.

To include the possibility of no waiting, we write $w_{n+1} = \max(w_n + s_n - t_n, 0)$. We wish to calculate the waiting-time distribution of a customer. Let us suppose that the $(n+1)$st customer will wait. We seek

$$\text{prob}(w_{n+1} < w) = \text{prob}(w_n + s_n - t_n < w) = \text{prob}(w_n < w - s_n + t_n).$$

Denote by $P_{n+1}(w)$ and $P_n(w)$ the distributions of the waiting times in the queue of the $(n+1)$st and nth unit. We assume that the variables t_n and s_n are independently distributed according to the distribution functions $A(x)$ and $B(y)$, and that the waiting time of the nth unit is independent of its own service time and of the interarrival time of the $(n+1)$st unit.

Accordingly, the probability that the waiting time w_n satisfies the above relation and that the interarrival time t_n and the service time s_n satisfy

$$x \leqslant t_n < x + dx, \quad y \leqslant s_n < y + dy,$$

is

$$P_{n+1}(w) = \int_0^\infty \int_0^\infty P_n(w + x - y)dA(x)dB(y).$$

This formula may also be obtained by considering the distribution of the sum of three independently distributed random variables. In the steady state the probabilities are independent of time and of the initial state of the system and all units have the same waiting time distribution $P(w)$. Hence

$$P(w) = \int_0^\infty \int_0^\infty P(w + x - y)dA(x)dB(y).$$

Example 2. Conflicting Populations

We now consider two populations in which a predator–prey situation exists. The rate of change of each population is assumed to be proportional to its present size, to the number of its encounters with the other population, and to a term which reflects the way in which the past history of the other population affects the present size of the given population. This yields the following equations for what has become known as a Volterra system.

$$\frac{dN_1}{dt} = aN_1 - bN_1N_2 - K_1N_1\int_0^t N_2(s)ds,$$

$$\frac{dN_2}{dt} = -cN_2 + dN_1N_2 + K_2N_2\int_0^t N_1(s)ds.$$

$N_1(t)$ and $N_2(t)$ are the numbers in populations 1 and 2; the remaining coefficients are positive constants.

*3.12. Stochastic Differential Equations

It is convenient to distinguish three basic types of stochastic differential equations. Such equations may involve either:

(i) random initial conditions;
(ii) random forcing functions; or
(iii) random coefficients.

In the first case, the given probabilistic properties of the initial conditions (regarded as random variables) determine those of the solution $x(t)$ through the explicit form of $x(t)$. A typical example is when $x(t)$ represents the motion of a particle subject to uncertainty only in its initial position and governed by deterministic forces. See Chapter 8 by R. Syski in Saaty, *Modern Nonlinear Equations.*

In the second case, the stochastic process representing random forcing functions is given, and the properties of the solution $x(t)$ are developed through probabilistic argument. A typical example arises in the study of the output of a physical system when the input is a Brownian-motion process.

Finally, in the third case, the parameters of the system are regarded as the random variables, and the stochastic properties of $x(t)$ reflect the influence of randomness imposed by the structure of the system. For example, the output of an electric circuit with randomly varying capacity is described by equations of this type.

We consider situations in which each type of equation might occur rather than taking very specific examples as in the earlier sections. Our examples may most easily be applied in dynamics, but there are many other problems where these conditions apply.

Example 1. Random Initial Conditions

Consider simple linear motion. This may be described by an equation of the form $\dfrac{dx(t)}{dt} = c$, $t \geqslant 0$, where $x(t) =$ position at time t and $c =$ constant. We assume $c \geqslant 0$.

The solution is given by

$$x(t) = ct + x_0,$$

where x_0 gives the position at time $t = 0$.

Suppose that x_0 is a random variable with a given distribution. The mean, $E\{x(t)\}$ and the variance, $\mathrm{var}\{x(t)\}$, of the random variable $x(t)$, can be calculated directly, even without the explicit form of this distribution. We obtain

$$E\{x(t)\} = ct + E\{x_0\}$$

$$\mathrm{var}\{x(t)\} = \mathrm{var}(ct + x_0) = \mathrm{var}\, x_0.$$

Suppose now that x_0 has a Gaussian distribution with mean zero and variance σ^2. Then the probability that x_0 assumes values between λ and $\lambda + d\lambda$ is given by $f_0(\lambda)d\lambda$, where the density $f_0(\lambda)$ has the form

$$f_0(\lambda) = \frac{1}{\sqrt{2\pi}\sigma} \exp\left(-\frac{\lambda^2}{2\sigma^2}\right), -\infty < \lambda < \infty.$$

We may now find the distribution of $x(t)$.

$$P\{x(t) < \mu\} = P\{ct + x_0 < \mu\} = P\{x_0 < \mu - ct\}$$

$$= \int_{-\infty}^{\mu - ct} f_0(\lambda) d\lambda = \int_{-\infty}^{\mu} f_0(\alpha - ct) d\alpha$$

$$= \int_{-\infty}^{\mu} f(\alpha, t) d\alpha.$$

Thus, $x(t)$ has a Gaussian density $f(\alpha, t)$, where

$$f(\alpha, t) = \frac{1}{\sqrt{2\pi}\,\sigma} \exp\left[-\frac{(\alpha - ct)^2}{2\sigma^2} \right],$$

which has the same variance σ^2 as x_0 but which has now a mean of ct.

Example 2. Random Forcing Functions

We are given a stochastic process (or a random function) $z = \{z(t), t \geqslant 0\}$ with specified stochastic properties. The differential equation, which we wish to consider, relates an unknown random function $x = \{x(t), t \geqslant 0\}$ and its derivatives to the given random function z. The solution of the equation will then define a new stochastic process $\{x(t), t \geqslant 0\}$ in terms of the given z process.

Because of the many physical applications, the function $z(t)$ is known as a *forcing function*.

To illustrate the problem, we consider the simple case of the first-order equation

$$\frac{dx(t)}{dt} + ax(t) = z(t) \quad (t \geqslant 0).$$

The solution is given by

$$x(t) = x(0)e^{-at} + \int_0^t e^{-a(t-\tau)} z(\tau) d\tau \quad (t \geqslant 0).$$

It can be seen that $x(t)$ depends on the values of $z(\tau)$ over the interval $0, t$, and it is not immediately clear whether the integral above defines a random variable. This brings up the question of what is meant by a *stochastic integral*.

A related problem is the meaning attached to the *stochastic* differentiability of $x(t)$; questions of existence and uniqueness of stochastic differential equations should then be examined. A point of view helpful in these considerations is to regard a random function, say x, as a function on the real line to the space of random variables. Thus, the value of x at t is a random variable $x(t, w)$, a point in a function space of random variables. Problems of continuity, differentiability, and integrability of random functions now become the same problems as those for ordinary functions which take values in abstract space; the same comment applies to differential equations.

Furthermore, explicit expressions for the distribution (or its transform) of $x(t)$ are required in terms of those for $z(t)$; this task, important for applications, is seldom easy. However, under suitable conditions, integration and averaging commute so that in the above example,

$$E\{x(t)\} = e^{-at}E\left[x(0)\right] \times \int_0^t e^{-a(t-\tau)}E\left[z(\tau)\right]d\tau.$$

Example 3. Random Coefficients

We consider the well-known application of damped simple harmonic motion

$$\ddot{x} + 2a\dot{x} + bx = 0.$$

Suppose that the coefficients a and b have a joint uniform distribution over the square $-1 \leqslant \alpha \leqslant 1$, $-1 \leqslant \beta \leqslant 1$, with density 1/4. We need to find the probability that every solution $x(t)$ tends to zero as $t \to \infty$ (i.e., the solution is stable). This occurs when zeros of the random characteristic equation

$$p^2 + 2ap + b = 0$$

are either (i) both real and both negative, or (ii) both complex with negative real parts. These two disjoint events may be described by random variables as follows. If $a^2 - b \geqslant 0$, then both roots are real, and they have the same sign if and only if $b > 0$, this sign being negative if and only if $a > 0$. If $a^2 - b < 0$, then the roots are complex conjugate, and they have negative real parts if and only if $a > 0$; in this case also $b > a^2 > 0$. Hence, the required event will occur if and only if $a > 0$ and $b > 0$. The probability that this will occur is:

$$P\left\{\lim_{t \to \infty} x(t) = 0\right\} = P\{a > 0, b > 0\} = \frac{1}{4}\int_0^1 \int_0^1 d\alpha \, d\beta = \frac{1}{4}.$$

Note also that the probability that the zeros of the characteristic equation are complex is

$$P\{a^2 - b < 0\} = \frac{1}{4}\int_{-1}^1 \int_{\alpha^2}^1 d\beta \, d\alpha = \frac{1}{3}.$$

3.13. Functional Equations

Generally, equations involving implicit relations between simple functions and functions of other functions are called functional equations. We give two simple examples.

Example 1. Total Purchase Expenditures

Consider a process which occurs in N stages. An example of this would be purchases over a period. Let x_n be the input at stage n, d_n be a decision or control variable at stage n (an example of a control variable is the amount purchased in a process at the nth stage), r_n be a return function which depends on x_n and d_n, and suppose that the input between two consecutive stages satisfies the relation

$$x_{n-1} = T_n(x_n, d_n).$$

Then the total purchase expenditure for (x_n) in n stages is given by an equation at the form

$$f_n(x_n) = \min_{d_n} r_n(d_n, x_n) + f_{n-1}(T_n d_n, x_n).$$

As $n \to \infty$ this equation becomes

$$f(x) = \min_d r(d, x) + f(Td, x),$$

which is a functional equation from which $f(x)$ must be determined.

*Example 2. Information Theory

From the standpoint of information or knowledge, the presence of a problem is a state of uncertainty. Solving the problem totally or partially removes or reduces this uncertainty. Thus, a problem may be first regarded in terms of the amount of uncertainty in it; logical operations and experimental tests may be used to gather information which decreases the uncertainty (to zero if possible). Uncertainty or entropy content (the negative of this quantity is defined as the information content) of a set of n mutually exclusive events is given by

$$H(p_1, \ldots, p_n) = - \sum_{i=1}^{n} p_i \log p_i \qquad \sum_{i=1}^{n} p_i = 1,$$

where p_i is the probability that the ith factor arises in defining the problem.

In the continuous case, the entropy is given by

$$H[f(x)] = - E[\log f(x)] = - \int_{-\infty}^{\infty} f(x) \log f(x) \, dx$$

$$\int_{-\infty}^{\infty} f(x) dx \equiv 1, \qquad \int_{-\infty}^{\infty} x f(x) dx = 1, \qquad \int_{-\infty}^{\infty} x^2 f(x) dx = \sigma^2.$$

The form which these expressions take arises from the requirement that the sum of the probabilities causing the occurrence of two independent events is equal to the probability of the occurrence of the two events together.

This yields a functional equation of the form

$$f(x) + f(y) = f(xy),$$

which has $f(x) = \log x$ as a solution. The expected value gives the entropy. Recall that

$$E[x] = \sum_{i=1}^{n} x_i f(x_i) \quad \text{or} \quad \int_{-\infty}^{\infty} x f(x) \, dx,$$

depending on whether x is discrete or continuous. Thus we obtain

$$\sum_{i=1}^{n} p_i \log p_i$$

or

$$\int_{-\infty}^{\infty} f(x) \log f(x) dx.$$

Chapter 3—Problems

1. Formulate the following problem algebraically. A band of 13 pirates obtained a certain number of gold coins. They tried to distribute them equitably but found that they had eight left. Two pirates died of smallpox. The remainder tried to distribute the coins equitably among the 11 remaining pirates, but they found that there were three left. Thereupon they shot three pirates, but still there was a remainder of five coins when they attempted an equal division among the eight remaining pirates. How many coins were there? (All numbers of the form $333 + 1144k$.)

2. What number, when added to 100 and to 164 yields the square of an integer? Find all solutions? (125, −64, −100.)

3. Consider a river in which the velocity of the current is V in the direction shown by the arrow (Fig. A). A man swims from A to B and back to A. Another man leaves A at the same time as the first, and, swimming at the same rate, swims to C and back to A. If the distances AB and AC are equal, AB being parallel to the current and AC perpendicular to it, which man gets back first? (*Hint*: Note that the man swimming towards C would have to aim for a point other than C, to allow for the effects of the current.)

FIG. A.

4. Suppose that the rate at which a rumor spreads varies directly with the number of people who have heard the rumor. Show that the rumor grows exponentially. Is this a good model? Is the assumption reasonable? Can you modify your equation to take care of possible problems?

5. Find P_n, the probability that an event will occur on the nth trial, given λ, the probability of its occurrence on the nth trial if it occurred on the $(n-1)$st trial and μ, the probability of its occurrence on the nth trial if it did not occur on the $(n-1)$st trial. $[p_n = \lambda p_{n-1} + \mu(1 - p_{n-1}).]$

References

Cascading cups, *Am. Math. Monthly*, Vol. 28, 1921, p. 144.

Edie, L. C., Car-following and steady state theory for non-congested traffic, *Operations Res.*, Vol. 9, No. 1, 1961, pp. 66–76.

Bibliography

For a comprehensive list of references on equations and inequalities, consult:

Saaty, Thomas L. and J. Bram, *Nonlinear Mathematics*, McGraw-Hill, New York, 1964.

Saaty, Thomas L., *Modern Nonlinear Equations*, McGraw-Hill, New York, 1967.

Saaty, Thomas L., *Elements of Queueing Theory, with Applications*, McGraw-Hill, New York, 1961.

Saaty, Thomas L., *Mathematical Methods of Operations Research*, McGraw-Hill, New York, 1959.

Saaty, Thomas L., *Optimization in Integers and Related Extremal Problems*, McGraw-Hill, New York, 1970.

Chapter 4

Optimization

4.1. Introduction

SOME of the most widely used models are those of maximization and minimization, simply called optimization. The pursuit of an optimum is a special case of a broader pursuit of an extremum. This may be a property which satisfies the vanishing of the first derivative or Euler's equation in the calculus of variations. More recently, solution of conflict problems—another area of optimization—has become a central preoccupation of game-theorists. As illustrated in the previous chapter, stochastic properties may be introduced into the formulation of an optimization problem.

In practicing optimization, one needs to keep in mind the fact that calculus methods apply only to a small class of problems; the majority have to be solved using ingenuity and good logical thinking. In recent years there has been a great surge of effort to expand and unify optimization methods. The unification falls into two areas: functional analysis and Diophantine or integer optimization.

An optimization problem consists of two parts. The first part is that a function (frequently called an objective function) has to be maximized or minimized. For example, the function may be a cost function whose coefficients are unit costs and whose variables are various quantities, such as foods to be purchased. The second part involves a set of constraints given in the form of equations or inequalities; the constraints express limitations on the variables which must be satisfied. An optimization problem need not have constraints.

The objective function may be linear in all its variables. Similarly, the constraints may be all linear. In that case, we have a linear-programming problem. In general the variables are constrained to be nonnegative, but this is not necessary. If any variable anywhere in the problem appears in other than a linear form, the problem becomes a nonlinear problem.

The constraints of an optimization problem define a region known as the feasible region from which the optimum of the objective function is to be found. In a linear-programming problem, the feasible region is convex–polyhedral in shape and the optimum is on the boundary, usually at a vertex of the polyhedron. If the problem is nonlinear, the optimum could be in the interior of the region.

The simplex process is an algorithm for solving a linear-programming problem. There are many algorithms for solving nonlinear programs. Dynamic programming is a successive iteration method of solving an optimization problem given in general functional form. In the calculus of variations, the objective function is usually an integral and the constraints may be algebraic or involve integrals and derivatives. In classical control theory, the constraints are differential equations representing a dynamic system. The objective function to be minimized here may be the time to achieve rendezvous by a rocket whose trajectory is described by the constraints.

We have discussed optimization problems in which there is a single objective function to be maximized or minimized. In many real-life problems in which there is a conflict of interest, game theory provides the rationale for formulating and solving such a problem. Even though game theory is illustrated in this chapter, most of the discussion on conflict resolution is postponed to the last chapter.

In addition to the specific references pertaining to the chapter, we have included a relatively comprehensive bibliography on optimization. We are grateful to our good friend, M. Z. Nashed, for his excellent contribution to the list.

4.2. Unconstrained Minima

Sometimes there are no constraints at all on the function to be optimized except perhaps for those implicit in its definition.

Example 1. Optimal Branching of Arteries [Rosen]

What is the optimal angle at which a branching from an artery takes place in order to reach a given position of the body? The size of the angle is a function of the amount of work (measured by the resistance of the system—the greater the resistance, the greater the work) to be done to push a given volume of fluid through the body (Fig. 4.1). Thus, lower resistance means less work by the heart. Resistance varies directly with the total length of the system (the greater the distance, the greater the resistance) and inversely with the radii of the vessels. Thus, too early a branching as the artery leaves the heart, implies a shorter length system, but also implies a decrease in the overall radius of the system since the cross-section is decreased by branching. Too late a branching (e.g., an angle perpendicular to the main vessel) implies a larger radius, but a larger system. The resistance R from the initial point of the main artery AB to the destination C is given by $R = R_1 l_1 + R_2 l_2$, where R_i is the resistance per unit length along the vessel whose length is $l_i, i = 1, 2$.

If l_0, θ and θ_0 are as shown in Fig. 4.1, then $R = R_1 l_1 + R_2 l_2 = R_1 l_0 (\cot \theta_0 - \cot \theta) + R_2 l_0 \operatorname{cosec} \theta$.

If r_1 is the radius of AD and r_2 is the radius of DC, then Poiseuille's law for fluid flow in rigid pipes gives $R_1 = k r_1^{-4}$, $R_2 = k r_2^{-4}$ and the expression for resistance R becomes:

$$\frac{k l_0 \cot \theta_0}{r_1^4} \text{ (constant)} + k l_0 \left[\frac{\operatorname{cosec} \theta}{r_2^4} - \frac{\cot \theta}{r_1^4} \right]$$

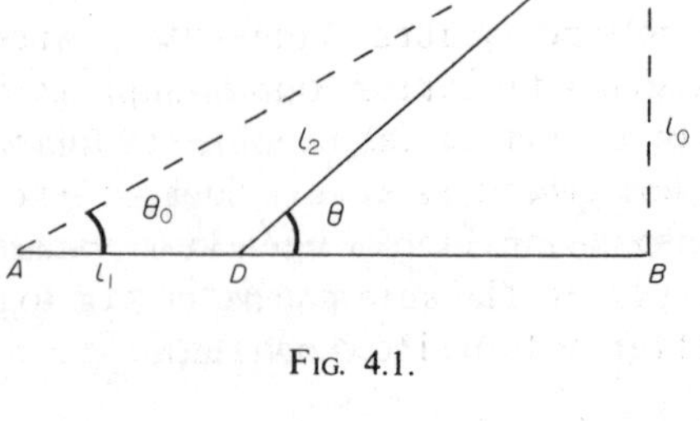

FIG. 4.1.

The minimum value of this expression with respect to θ occurs when

$$\theta = \text{arc } \cos(r_2^4/r_1^4).$$

To bring the result closer to experimental observation, we include a factor concerning the minimum amount of work done to maintain the system. This work may be assumed to be proportional to the volume of the system which, in this case, is $l_1\pi r_1^2 + l_2\pi r_2^2$. If we use K as the constant of proportionality, the minimum angle is now

$$\theta = \frac{k + Kr_1^6}{k + Kr_2^6} \text{ arc } \cos(r_2^4/r_1^4).$$

As $r_2 \to r_1$ the two results become the same.

Example 2. A Location Problem

It is desired to find the location (u, v) of an electrical utility plant to supply the needs of users with locations at $(x_i, y_i) i = 1, 2, \ldots, N$ and minimize the transmission losses. The square of the distances r_i, $i = 1, 2, \ldots, N$ from the location is given by

$$r_i^2 = (x_i - u)^2 + (y_i - v)^2.$$

Let the concentration of users at location i be n_i and assume that the total loss function to be minimized takes the form

$$L = n_1 g(r_1) + n_2 g(r_2) + \ldots + n_N g(r_N),$$

where $g(r_i)$ is the loss associated with the distance r_i.

The problem now is to find u and v to minimize L. We assume that the function $g(r)$ has the form $g(r) = kr^2$, where k is a constant. (This is a reasonable approximation in the case of electric cables.) We substitute for $g(r_i)$ and r_i in L and differentiate with respect to u and v. Equating these partial derivatives to zero and solving yields the physical center of gravity of the N locations:

$$u = \sum_{i=1}^{N} n_i x_i \Big/ \sum_{i=1}^{N} n_i, \quad v = \sum_{i=1}^{N} n_i y_i \Big/ \sum_{i=1}^{N} n_i.$$

Consider now what happens if we let

$$L = \sum_{i=1}^{N} n_i s_i \{(x_i - u)^2 + (y_i - v)^2\}^{1/2},$$

where s_i, $i = 1, 2, \ldots, N$, is the amount of electricity used by the "average" user at location i. If we put $n_i s_i = w_i$ and interpret w_i as a weight, then $\sum_{i=1}^{N} w_i r_i$ is the sum of products of weights and distances and may be regarded as the potential energy of a physical system. Recall that any physical system adjusts itself so as to minimize its potential energy. The equations resulting from differentiation can be solved as follows.

Form a plan of the area concerned on a wooden table, bore holes at the N location and place a series of weights proportional to the w_i on long strings of equal length passing through the corresponding holes. The strings are then tied together on the table. The system of strings and weights is disturbed to allow it to move to a position of equilibrium.

When it comes to rest, the position of the knot will indicate the best location of the utility plant.

*4.3.   Constrained Problems

4.3.1.   LAGRANGE MULTIPLIERS

One is often faced with the problem of finding the optimum while satisfying a number of conditions or constraints. Consider the following general problem: One wishes to find an extremum of a differentiable function $f(x_1, \ldots, x_n)$ whose variables are subject to the m constraints

$$g_i(x_1, \ldots, x_n) = 0 \quad (i = 1, \ldots, m; \quad m \leqslant n),$$

where the g_i are also differentiable.

Form the Lagrangian function

$$F(x_1, \ldots, x_n; \lambda_1, \ldots, \lambda_m) = f(x_1, \ldots, x_n) + \sum_{i=1}^{m} \lambda_i g_i(x_1, \ldots, x_n),$$

involving the Lagrange multipliers $\lambda_1, \ldots, \lambda_m$. Then the necessary conditions for an unconstrained extremum of F (namely, the vanishing of the first partial derivatives) are also necessary conditions for a constrained extremum of $f(x_1, \ldots, x_n)$, provided that the matrix of partial derivatives $\partial g_i / \partial x_j$ has rank m at the point in question. These necessary conditions are a system of $m + n$ equations

$$\frac{\partial F}{\partial x_j} = \frac{\partial f}{\partial x_j} + \sum_{i=1}^{m} \lambda_i \frac{\partial g_i}{\partial x_j} = 0 \quad (j = 1, \ldots, n),$$

$$\frac{\partial F}{\partial \lambda_i} = g_i = 0,$$

which one can then (at least in theory) solve for the $m + n$ unknowns $x_1, \ldots, x_n$; $\lambda_1, \ldots, \lambda_m$. (One is really interested only in obtaining $x_1, \ldots, x_n$.)

Example 1. Eat Some of Every Food

Let the rate of change of the utility of a quantity x with respect to x be inversely proportional to the amount of the quantity already available. Thus if $u(x)$ is the utility of x, we have

$$\frac{du}{dx} = \frac{a}{x} \quad \text{from which} \quad u(x) = a \log x,$$

i.e., the more we have the smaller is the marginal utility obtained from it.

Consider now the utility function $u(x_1, \ldots, x_n)$, where $x_i, i = 1, 2, \ldots, n$ is the amount of food i. Suppose $u(x_1, \ldots, x_n) = \sum_{i=1}^{n} a_i \log x_i$, where a_i is the amount of pleasure obtained from the ith food. It is desired to maximize $u(x_1, \ldots, x_n)$ subject to $\sum_{i=1}^{n} P_i x_i \leqslant M$ where P_i

is the number of calories in a unit of food i and M is a bound on the total caloric intake. The solution to this problem (using Lagrange multipliers) is given by $x_i = a_i M / P_i \sum\limits_{i=1}^{n} a_i$. Thus, to maximize pleasure, one must eat some from each kind of food (assuming that each $a_i > 0$), contrary to certain dietary suggestions that some foods must be cut out completely.

Example 2. Optimum Dimensions of a Multilevel City

A problem which arises in a three-dimensional (square prism) city planning design is concerned with finding the dimensions which would minimize the maximum possible travel time between two diametrically opposite farthest points, subject to a given floor area of the building. Each floor is assumed to have a square grid of streets. If we assume that vertical elevator speed is α and horizontal speed is β, then the problem is to minimize the time to travel the height of the city h and twice the side k of its square base. This leads to the expression for the time T,

$$T = \frac{h}{\alpha} + \frac{2k}{\beta}.$$

Since the total internal floor area A is given, we have

$$nk^2 = A,$$

where n is the number of levels above the ground. Assuming that c is the given height of each floor, then

$$n = \frac{h}{c}.$$

The optimization problem then is to minimize T subject to the constraint

$$\left(\frac{h}{c}\right) k^2 = A.$$

Using the Lagrange multiplier method, we obtain the following:

$$h = \left(\frac{Ac\,\alpha^2}{\beta^2}\right)^{1/3}, \quad k = \left(\frac{c\beta A}{\alpha}\right)^{1/3}.$$

*Example 3. A Walk in the Rain [Deakin]

A man wishes to walk (without an umbrella) from A to B (vector direction $\mathbf{i}$) while it is raining (unit downward direction $-\mathbf{k}$ and terminal rain speed V_t). Let $\mathbf{j} = \mathbf{i} \times \mathbf{k}$ (cross-product orthogonal to the $\mathbf{i}$, $\mathbf{k}$ plane). There is a horizontal wind of velocity $V_t(w\mathbf{i} + W\mathbf{j})$. Thus, the rain velocity is $V = V_t(w\mathbf{i} + W\mathbf{j} - \mathbf{k})$. The man's speed is $u - xV_t$. Hence, the velocity of the rain with respect to the man is $V_t[(w - x)\mathbf{i} + W\mathbf{j} - \mathbf{k}]$. Consider the man as a rectangular prism with three sides (top with area αA, right or left with area βA and front or back with area A) getting wet. What should his optimal speed be? The amount of wetness per unit time is proportional to the cosine of the angle between the normal to the side the rain hits and the relative rain direction multiplied by the area of the side and is inversely proportional to his speed. Since the normals are the unit vectors $\mathbf{i}, \mathbf{j}, \mathbf{k}$, the total amount of rain falling on the man is proportional to (using the cosines of the angles between the direction line of the rain and the top, right front, or back, respectively and using the speed $x V_t$):

$$\overset{\text{Top}}{} \quad \overset{\text{Right or left}}{} \quad \overset{\text{Front or back}}{} \quad \overset{\text{Speed factor}}{}$$
$$F(x) = (\alpha A + \beta A |W| + A|w - x|) \Big/ x$$

It is desired to minimize the "wetness function" $F(x)$. The absolute values arise from the fact that just two of the four vertical sides can be wet depending on the signs of W and $(x - w)$.

The optimal policy for walking in the rain depends on whether $\alpha + \beta |W| > w$ or not. In the former case $F(x)$ is monotone decreasing for all x and the optimal strategy is to run as fast as possible. Otherwise, $F(x)$ has a minimum at $x - w$ and the man should travel at the speed $x = w$ if he can to avoid wetness on front or back. He will get wet only on the top and side. If he cannot attain w, he should run as fast as he can. We assume that $\alpha = 0.06$, $\beta = 0.33$, $V_t = 20$ mph. Let X be his top speed, and let $\alpha + \beta W < w$. Then he should run as fast as w or x. This question arises only if $X > w$. If we measure relative wetness by $R = [F(x)]/[F(w)]$, R is maximized when $\alpha + \beta |W|$ is minimized at α; i.e., when $W = 0$. This leads to

$$R_{max} = \frac{(X + x)^2}{4X \alpha} = 4.68,$$

where $x = 1$. This means he gets more than four times wetter than if he jogged with the same speed as the wind. Note that this refers only to the *rate* of wetness. The total wetness will also depend on time.

*4.3.2. UNKNOWN NUMBER OF VARIABLES

Most optimization problems deal with a fixed number of variables. The following is an example with the number of variables to be determined together with their optimal values.

Example. The Jeep Problem (optimization with constraints)

We shall illustrate a problem in which it is desired to minimize a function subject to constraints given as recursive relations.

Problem. It is desired to advance a vehicle 1000 miles from an original position with a minimum expenditure of fuel. The vehicle has maximum fuel capacity of 500 units which it uses at the uniform rate of one unit per mile. The vehicle must stock its own storage points from its one tank by making round trips between storage points. Determine the storage points along the route which minimize the fuel consumption for the entire trip. Determine the number of round trips required between each pair of storage points. Determine the minimum amount of fuel required at the start.

If s_i is the amount of fuel stored at the ith storage point $(i = 0, 1, \ldots, n)$, d_i is the distance between the $(i - 1)$st and the ith storage points, and k_i is the number of round trips the vehicle makes between these two points, then

$$s_{i-1} = s_i + 2k_i d_i + d_i \quad (i = 1, \ldots, n). \tag{4.1}$$

Thus, the amount of fuel stored at the $(i-1)$st point is equal to the amount stored at the ith point plus the amount consumed along the route in making k_i round trips and a single forward trip.

It is not difficult to see that a minimum use of fuel is made if for each trip the vehicle proceeds loaded at full capacity. For, in that case, a smaller number of trips would be made and, hence, less fuel is consumed in travel. Also, in order to have no fuel left at the end, it is necessary that the vehicle be loaded with 500 units of fuel at the 500-mile point. From these two facts, it follows that the ith storage point should be located so that the vehicle makes k_{i+1} round trips between this point and the $(i+1)$st storage point and one last forward trip, always fully loaded and ultimately leaving no fuel behind at the ith storage point. Working backward, the last statement is valid back to the first storage point but is not possible for the starting point since the position of that point is predetermined. Hence, the vehicle will make its last forward trip between the starting point and the first storage point with a load $c < 500$. Thus, we have:

$$\left.\begin{aligned} s_i &= k_i(500 - 2d_i) + 500 - d_i \quad (i = 2, \ldots, n), \\ s_1 &= k_1(500 - 2d_1) + c - d_1. \end{aligned}\right\} \tag{4.2}$$

It is desired to minimize s_0 given by

$$s_0 = s_1 + 2k_1 d_1 + d_1.$$

Now, using for s_1 the value previously given, one has

$$s_0 = 500k_1 + c. \tag{4.3}$$

Since the vehicle can travel the last 500 miles without need for stored fuel, in order to minimize s_0, it suffices to put $s_n = 500$ and to place the storage points along the first 500 miles of the route. Thus,

$$\sum_{i=1}^{n} d_i = 500.$$

It turns out that $k_i = 8 - i$, $d_i = 500/(17 - 2i)$, $i = 1, \ldots, 7$ and $s_0 = 3836.45$.

4.4. Linear Programming

One of the most important special cases of optimization deals with linear objective functions subject to linear inequality constraints. This problem may be stated mathematically as follows: find values $(x_1 \ldots x_n)$ which maximize the linear form

$$x_1 c_1 + \ldots + x_n c_n$$

subject to the conditions

$$x_j \geqslant 0 \quad (j = 1 \ldots n)$$

and

$$\sum_{j=1}^{n} a_{ij} x_j \leqslant b_i \quad (i = 1 \ldots m),$$

where the a_{ij}, b_i, c_j are constants.

In matrix notation the linear-programming problem may be stated as follows: Find $\mathbf{x}$ to maximize $\mathbf{c}'\mathbf{x}$ subject to $\mathbf{x} \geqslant 0$ and $A\mathbf{x} \leqslant \mathbf{b}$, where, if

$$\mathbf{c} = \begin{bmatrix} c_1 \\ \ldots \\ c_n \end{bmatrix} \quad \text{then} \quad \mathbf{c}' = [c_1 \ldots c_n]$$

is its transpose, and

$$\mathbf{b} = \begin{bmatrix} b_1 \\ \ldots \\ b_n \end{bmatrix}$$

$$A = \begin{bmatrix} a_{11} a_{12} \ldots a_{1n} \\ a_{21} a_{22} \ldots a_{2n} \\ \ldots\ldots\ldots \\ a_{m1} a_{m2} \ldots a_{mn} \end{bmatrix}, \quad \mathbf{x} = \begin{bmatrix} x_1 \\ x_2 \\ \vdots \\ x_n \end{bmatrix}$$

Note that maximizing $\mathbf{c}'\mathbf{x}$ is equivalent to minimizing $-\mathbf{c}'\mathbf{x}$, and vice versa.

The following two examples illustrate problems which can be solved using linear-programming methods.

Example 1. The Diet Problem

Suppose that a thrifty patient of limited means was advised by his doctor to increase the consumption of liver and frankfurters in his diet. In each meal he must get no less than 200 calories from this combination and no more than 14 units of fat. When he consulted his diet book, he found the following information: there are 150 calories in a pound of liver and 200 calories in a pound of frankfurters. However, there are 14 units of fat in a pound of liver and four units in a pound of frankfurters.

So, he reasoned as follows: I must eat an amount x_1 of liver and an amount x_2 of frankfurters, paying the least amount of money (the price is 30 cents per pound of liver and 50 cents per pound of frankfurters) and meeting the requirements set by the doctor. There are $150x_1$ calories in x_1 pounds of liver and $200x_2$ in a pound of frankfurters; the total amount of calories obtained, that is, $150x_1 + 200x_2$, must be no less than 200, and so one writes

$$150x_1 + 200x_2 \geqslant 200.$$

Similarly, the total fat should be no greater than 14 units:

$$14x_1 + 4x_2 \leqslant 14,$$

and the total cost $30x_1 + 50x_2$ must be a minimum. He is faced with minimizing the total cost subject to the medical constraints (hence the term *constrained optimization*). He must eat something but just enough to meet the requirements and at minimum cost.

This problem can be solved in several ways. For a more complicated problem with many more dietary items which the patient must eat and with further restrictions such as the amount of vitamins obtained, one uses the well-known simplex process for its solution. For the simple problem given above, one can draw a picture (Fig. 4.2) and find the desired solution; i.e., the food quantities x_1 and x_2 to satisfy the constraining inequalities and yield the minimum cost.

To plot the inequalities, one simply plots a straight line, using only the equality sign (as in Fig. 4.2).

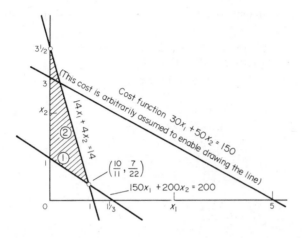

FIG. 4.2.

Then the first inequality is satisfied by all the points to the right of the line and the second inequality by the points to the left of the corresponding line. Thus the points common to both inequalities lie in the shaded region. The cost expression is given an arbitrary value at first. Note, however, that it may be moved back and forth in a parallel direction. The minimum is obviously given when the line is moved as close to the origin as possible, while still intersecting the shaded region. This obviously occurs at $x_1 = \frac{10}{11}$, $y_1 = \frac{7}{22}$, as indicated by the arrow. Consequently, this value, which is roughly one pound of liver and a third of a pound of frankfurters, satisfies the requirements at minimum cost.

Example 2. The Caterer Problem [Jacobs]

A caterer knows that, in connection with the meals he has arranged to serve during the next n days, he will need $r_j \geqslant 0$ fresh napkins on the jth day, $j = 1, 2, \ldots, n$. Laundering normally takes p days; i.e., a soiled napkin sent for laundering immediately after use on the jth day is returned in time to be used again on the $(j + p)$th day. However, the laundry also has a higher-cost service which returns the napkins in $q < p$ days (p and q are integers).

If the caterer has no clean napkins on hand, he will meet his immediate needs by purchasing napkins at a cents each. Laundering costs b and c cents per napkin for the normal and high-cost service respectively. How does he arrange matters to meet his needs and minimize his outlay for the n days?

Let x_j represent the napkins purchased for use on the jth day; the remaining requirements, if any, are supplied by laundered napkins. Of the r_j napkins which have been used on that day plus any other soiled napkins on hand, let y_j be the number sent for laundering under normal service and z_j the number under the rapid service. Finally, since soiled napkins need not be sent to the laundry immediately after use, let s_j be the stock of soiled napkins on hand after $y_j + z_j$ have been shipped to the laundry; they will be available for laundering on the next day, together with those from the r_{j+1} used on that day.

Consequently,

$$y_j + z_j + s_j - s_{j-1} = r_j \quad (j = 1, 2, \ldots, n).$$

That is, the napkins used on the jth day are equal to the number sent to the laundry plus the change in the stock of soiled napkins.

The returns from the laundry are equal to the amounts shipped for normal service p days earlier plus the amounts shipped for rapid service q days earlier. Together with purchases, these provide for the needs on the same day. This gives

$$x_j + y_{j-p} + z_{j-q} = r_j \quad (j = 1, 2, \ldots n).$$

The total cost to be minimized, subject to these constraints, is

$$\sum_j (ax_j + by_j + cz_j),$$

where

$$a > b, \quad c > b; \qquad x_j, \quad y_j, \quad z_j, \quad s_j \geqslant 0.$$

4.5.   Nonlinear Programming

If we remove the hypotheses of linearity from a linear programming problem, we get a nonlinear programming problem.

The general programming problem has the form

$$\text{maximize } f(x_1 \ldots x_n)$$

subject to the constraints

$$g_i(x_1 \ldots x_n) \leqslant 0 \quad (i = 1 \ldots m),$$
$$x_j \geqslant 0 \quad \text{or} \quad g_{m+j} \equiv -x_j \leqslant 0 \quad (j = 1 \ldots n).$$

Minimizing f is equivalent to maximizing $-f$.

Examples of nonlinear programming problems abound in the literature. The following are among the better known problems.

Example 1. **Maximizing Revenue**

A monopolist wishes to maximize his revenue. The sale price p_j of a commodity depends on the quantity of each of several commodities which he produces (unlike the case of free competition). Thus, if the revenue is given by $f(x) = \sum_{j=1}^{n} p_j x_j$, where p_j is the price of the jth commodity and x_j is the amount of the jth commodity to be produced, then

$$p_i = b_i - \sum_{j=1}^{n} a_{ij} x_j \quad (i = 1 \ldots n).$$

If we substitute these quantities in f, we obtain a quadratic function to be maximized subject to nonnegativity constraints $x_j \geqslant 0$, $j = 1 \ldots n$, and subject to capacity constraints which may be prescribed in a linear form. Thus, since $f(x)$ is nonlinear, we have a nonlinear program.

A similar problem may be formulated in which the cost of producing a commodity depends on the quantity produced.

Example 2. Commodity Production

In this case the objective is to find the values of x_j which minimize the quadratic function $f(x)$ subject to $x_j \geq 0$ and subject to the constraints that the entire capacity should be used. These constraints can be expressed in terms of linear inequalities with nonnegative coefficients.

Example 3. Portfolio Selection

Yet another example is that of portfolio selection in which an investor wishes to allocate a given sum of money C to different securities whose total return is considered a random variable. He may either maximize the expected return for a fixed variance or minimize the variance for a fixed return. If $x_j, j = 1 \ldots n$, is the amount to be invested on the jth security, then $\sum_{j=1}^{n} x_j = C$. The total return is $\sum_{j=1}^{n} r_j x_j$, its expected value is $E = \sum_{j=1}^{n} \mu_j x_j$, and its variance is $\sum_{i,j=1}^{n} a_{ij} x_i x_j$, where (a_{ij}) is the covariance matrix of the random variables r_j, which represent the return on the jth security, and μ_j is the expected value of n_j. Both (a_{ij}) and μ_j are given. Here the problem is either to maximize $\sum_{j=1}^{n} \mu_j x_j$ subject to $\sum_{ij} a_{ij} x_i x_j \leq \sigma$, where σ is a given constant, and subject to $\sum_{j=1}^{n} x = C$, $x_i \geq 0$, or to minimize $\sum_{ij} a_{ij} x_i x_j$ subject to $\sum_{j=1}^{n} \mu_j x_j = \mu$, where μ is given, and subject to $\sum_{j=1}^{n} x_j = C$, $x_j \geq 0$. These problems are obviously nonlinear, in the first case because of the presence of a quadratic constraint, and in the second because this constraint is now the function to be minimized.

Example 4. Chemical Equilibrium

Next, we consider an example in chemical equilibrium. We are told that m gaseous elements are used to form n compounds under constant pressure P, where there are a_{ij} atoms of the ith element in one molecule of the jth compound. If there were b_i moles of the ith element used and there are now x_j moles of the jth compound in the mixture, then, by conservation of mass,

$$\sum_{j=1}^{n} a_{ij} x_j = b_i \quad (i = 1 \ldots m)$$

with

$$x_j \geq 0 \quad (j = 1 \ldots n)$$

with $n \geq m$.

Subject to these constraints it is desired to minimize the total Gibbs free energy of the system:

$$\sum_{j=1}^{n} c_j x_j + \sum_{j=1}^{n} x_j \log \frac{x_j}{\sum_{i=1}^{n} x_i},$$

where $c_j = F_j/RT + \log P$, F_j is the Gibbs energy per mole of jth gas at temperature T and unit atmospheric pressure, and R is the universal gas constant.

A linear-programming problem in which the solution must be integral is called an integer-programming problem. Sometimes the solution must be integral even for a nonlinear problem.

Example 5. Nonlinear Programming in Integers

Given a system of m stages which are connected in series; the ith stage consists of n_i components connected in parallel as indicated in Fig. 4.3.

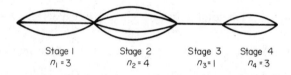

Stage 1	Stage 2	Stage 3	Stage 4
$n_1 = 3$	$n_2 = 4$	$n_3 = 1$	$n_4 = 3$

FIG. 4.3.

The probability that an ith-stage component will function is p_i; the cost of the component is c_i. The system can only function if at least one component in each stage is operable. We are concerned with the maximum reliability of the system and the minimum cost of the system.

Since the probability that a component of the ith stage is operable is p_i, $1 - p_i$ is the probability that it is inoperable, $(1 - p_i)^{n_i}$ is the probability that none of the components of the ith stage is operable, and $1 - (1 - p_i)^{n_i}$ is the probability that at least one component of the ith stage is operable. Since the stages are in series we take the product of these probabilities to obtain the probability $P(n_1 \ldots n_m)$ that the system is operable. Thus

$$P(n_1 \ldots n_m) = \prod_{i=1}^{m} \{1 - (1 - p_i)^{n_i}\}.$$

The cost of the system is given by $C = \sum_{i=1}^{m} c_i n_i$. The problem may be either to maximize P with respect to integral values of n_i for a given value of C or to minimize C for a given value of P.

4.6.   Multistage Optimization (Dynamic Programming)

The dynamic programming technique enables one to handle a multistage decision process as a series of single problems.

The process may be defined as a transformation, made up of a series of independent stages, each of which is characterized by a set of parameters called "state variables." Dynamic-programming problems often use stages to represent intervals of time or space. Each time a new stage is reached, a choice among possible decisions must be made based on the values of the state variables at that stage. This choice will optimize the function over all stages already considered.

At each stage, the principle of optimality is assumed to hold; an optimal set of decisions has the property that whatever the first decision is, the remaining decisions must be optimal with respect to the outcome which results from the first decision. As a result, the transformation preserves the optimality of the original objective function.

We give a general example which shows how the method may be used.

Example. Buying and Selling

A merchant wants to plan his purchases and sales for a year. Let the prices for purchasing and selling in month i be p_i and c_i, respectively, after numbering the months backwards (i.e., Jan. $= 12$, Feb. $= 11, \ldots$). The goods are received at the beginning of each month and sold by the end of that month (Fig. 4.4). The storage capacity available is S. Find the optimal plan for the merchant.

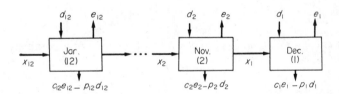

Fig. 4.4.

Let x_i be the amount of goods already at the store at the beginning of month i, d_i be the amount bought at the beginning of month i, e_i be the amount sold at the end of month i, $f_i(x_i)$ optimal profit feasible from month 1 to month i, as a function of x_i.

We have

$$f_1(x_1) = \max[c_1 e_1 - p_1 d_1],$$

where

$$0 \leq d_1 \leq S - x_1$$

and

$$0 \leq e_1 \leq x_1 + d_1,$$

$$f_i(x_i) = \max[c_i e_i - p_i d_i + f_{i-1}(x_i + d_i - e_i)]$$

$$i = 2, 3, 4, \ldots, 12,$$

$$0 \leq d_i \leq S - x_i,$$

$$0 \leq e_i \leq x_i + d_i.$$

Solving these equations recursively and computing d_i, e_i for $i = 1, 2, \ldots, 12$ makes it possible to evaluate the maximal profit $f_{12}(x_{12})$.

*4.7. Calculus of Variations

The calculus of variations is concerned with looking for a function which optimizes an integral subject to constraints generally given in algebraic form or involving integrals.

A relatively general problem in the calculus of variations is the problem of Bolza formulated by Bliss.

In a class of arcs

$$y(x) = [y_1(x) \ldots y_n(x)], \quad x_1 \leqslant x \leqslant x_2$$

find one which minimizes

$$J = F[x_1, y(x_1), \quad x_2, y(x_2)] + \int_{x_1}^{x_2} f(x, y, y')dx$$

subject to differential-equation constraints

$$g_i(x, y, y') = 0 \quad (i = 1 \ldots m < n)$$

and to the end conditions

$$h_i[x_1, y(x_1), x_2, y(x_2)] = 0 \quad (i = 1 \ldots p \leqslant 2n + 2)$$

(these conditions may separate into those exclusively on x_1 and those on x_2), where

$$y'(x) = [y'_1(x) \ldots y'_n(x)] = \left(\frac{dy_1}{dx} \ldots \frac{dy_n}{dx}\right)$$

Note that the end points x_1 and x_2 may vary as functions of a parameter vector

$$t = (t_1 \ldots t_r),$$

which in turn implies that

$$y_k(x_1) \equiv y_{k1}(t), \quad y_k(x_2) \equiv y_{k2}(t) \quad (k = 1 \ldots n)$$

and J becomes

$$J = F(t) + \int_{x_1(t)}^{x_2(t)} f(x, y, y')dx.$$

Two problems due to Mayer and Lagrange with variable end points were defined, the first requiring $F \equiv 0$ and the second requiring $f \equiv 0$, respectively.

A variational problem may involve retarded arguments such as in

$$\text{optimize} \int_{t_0}^{t_1} F[t, x(t), x(t-\tau), \dot{x}(t), \dot{x}(t-\tau)]dt$$

subject to the conditions

$$\dot{x}(t) = \phi(t) \quad \text{for} \quad t_0 - t \leqslant t \leqslant t_0$$

and $x(t_1) = x_1$; τ is a positive constant of retardation.

An auxiliary condition as constraint may be included such as

$$\int_{t_0}^{t_1} G[t, x(t), x(t-\tau), \dot{x}(t), \dot{x}(t-\tau)] dt = C,$$

where C is a given constant.

A very general variational problem can be formulated in the form of functional expressions such as:

$$\text{optimize } F[f(x, y, y')]$$

subject to differential, integral, difference, and a mixture of equality and inequality constraints. Two very practical examples follow.

Example 1. A Search Problem [Koopman]

The following variational example involves inequality constraints but is solved by elementary methods. Assume that $p(x, y) \, dx \, dy$ is the probability of the existence of a target in an area element $dx \, dy$ of a region R and that $e^{-\phi(x, y)}$ is the probability of not detecting it at a point in that region, where $\phi \geq 0$ and

$$\int_R \phi(x, y) = C$$

is the total effort available for search in the region. Find

$$\min_{\phi} \int_R p(x, y) e^{-\phi(x, y)} \, dx \, dy \equiv F[\phi]$$

subject to the above constraints. $F(\phi)$ is, of course, the probability of not detecting the target in R, if it is there. Note that this problem has a solution if $p(x, y)$ and $\phi(x, y)$ are integrable.

Example 2. Production Scheduling [Modigliani and Holm]

Find the most profitable schedule of production of a commodity over a given period of time $(0, T)$ so as to meet the following requirements:

1. The initial inventory h_0 is given.
2. The sales schedule is given, $S(t)$ being the cumulative sales from 0 to t (dS/dt need not be continuous).
3. Inventory can never be negative.
4. The terminal inventory is 0.
5. The cost of production is given: let $f(x)$ be the cost, per unit time, of producing x units of product per unit time; $f(x)/x$ is then the average cost, and $f'(x)$ is the marginal cost which is increasing.
6. The cost of storage is α per unit product and per unit time.

We define $X(t) = h_0 + \text{cumulative output up to time } t$. This ensures continuity for $X(t)$

but not for $X'(t)$. No restrictions on $X(t)$ other than the above are imposed. The assumptions are that the sales function $S(t)$ is any piecewise continuous function and the marginal cost function is monotone increasing and differentiable. The problem then becomes:

1. $X(0) = h_0$, the initial inventory.
2. $X(T) = S(T)$, in order to meet demand.
3. $X(t) \geqslant S(t)$ for $0 \leqslant t \leqslant T$ since the inventory cannot be negative.
4. $X(t)$ is continuous and nondecreasing and has a piecewise continuous derivative.

The cost

$$C = \int_0^T [\alpha(X - S) + f(X'(t))]dt,$$

where the first term of the integrand is storage cost per unit at time t and the second term is the production cost per unit time at time t.

Determine $X(t)$ which minimizes the total cost C subject to conditions 1 to 4.

4.8.   Theory of Optimal Control

A major problem of control theory is to find the "control vector" $y(t)$ which minimizes a scalar function

$$\int_0^T f(x, y, t)\, dt$$

subject to the vector differential equation

$$\frac{dx}{dt} = F(x, y), \quad x(0) = C$$

to local constraints

$$h_i(x, y) \leqslant 0 \quad (i = 1 \ldots m),$$

to global constraints

$$\int_0^T f_i(x, y)\, dt \leqslant a_i,$$

and to terminal conditions of the form

$$g_i(x(T), y(T), T) \leqslant 0.$$

The minimization may be alternatively applied to an expression in which T depends on the history of the process, i.e., $T = T(x, y)$. A number of applications of control theory have been extended to other areas. For example, in the field of arms control one may take Richardson's equations (see page 142 for details) with control $u_i(t) \quad i = 1, 2,$

$$\frac{dx}{dt} = ky - ax + g + u_1(t),$$

$$\frac{dy}{dt} = lx - by + h + u_2(t),$$

as constraints and find u_1 and u_2 which minimize the cost function, as armament given by

$$\int_0^1 (a_1 x^2 + a_2 y^2 + b_1 u_1^2 + b_2 u_2^2)\, dt,$$

where the coefficients a_i, b_i are given weights.

An example in economics follows.

Example 1. Saving and Investment

Let $S(t)$ be saving accumulated to time t, $I(f)$ the income, $C(t)$ the consumption, $R(t)$ the investment return all at time t ($R = aS$ where a is the compound interest rate per annum); then

$$\frac{dS}{dt} = I + aS - C$$

and if $S(0) = S_0$, then for any t in the interval $(0, T)$, we have

$$S(t) = S_0 e^{at} + \int_0^t e^{-ax}[I(x) - C(x)]\, dx.$$

The problem is to find $C(t)$ which maximizes some function

$$\int_0^T f[t, C(t)]\, dt$$

subject to the above equation with $t = T$ as constraint with the assumption that $I(f)$, $S(T)$, and $S(0)$ are known. A suitable choice of function f would be

$$f[t, C(t)] = e^{-bt} \log[1 + C(t)].$$

Example 2. An Advertising Model [Vidale and Wolfe]

A firm produces a single product and sells it in a market which can absorb no more than μ dollars of the product per time unit.

It is assumed that if the firm does not advertise at a certain time, its rate of sales will decrease in proportion to the sales at that time. If, on the other hand, the firm advertises, the rate of sales will increase proportional to the level of advertising, but this increase affects only that share of the market which is not already purchasing the product.

The problem is to find the optimal advertising policy, which maximizes total sales in a given time interval. Let $S(t)$ be the amount of sales at time t in dollars, $A(t)$ be the amount of advertising at time t in dollars, and λ, γ be positive constants.

The problem thus may be formulated as follows. Find

$$\max_{\substack{A(t)\text{ for}\\ t\varepsilon[0,\,T]}} \int_0^T S(t)\, dt$$

subject to

$$\dot{S}(t) = \frac{dS(t)}{dt} = -\lambda S(t) + \gamma A(t)\left[1 - \frac{S(t)}{\mu}\right]; \quad S(0) = S_0; \quad 0 \leqslant A(t) \leqslant \overline{A}$$

for some given constant $\overline{A}$.

Defining a new variable

$$Z(T) = \int_0^T S(t)\, dt$$

we obtain the following:

$$\max_{\substack{A(t)\ \text{for} \\ t \in [0,\,T]}} Z(T)$$

subject to:

$$\dot{S}(t) = -\left[\lambda + \frac{\gamma}{\mu} A(t)\right] S(t) + \gamma A(t); \quad S(0) = S_0$$

$$\dot{Z}(t) = S(t); \qquad\qquad\qquad\qquad Z(0) = 0.$$

The optimal policy turns out to be what is known as a "bang-bang" policy (i.e., advertising takes place in spurts, not continuously). The problem may be solved in its first formulation by means of the calculus of variations (Lagrange's problem), but this becomes a lengthy method. Euler's equation gives a nonlinear integro-differential equation, which can be solved only by numerical techniques.

4.9. Stochastic Optimization

A number of optimization problems involve stochastic elements and require a special treatment. The stock-piling example treated above is an illustration. A typical problem may involve the maximization of a decision function based on the variables of another optimization problem. Thus, one seeks to maximize a function

$$f(x_1 \ldots x_n;\ c_1 \ldots c_r),$$

where the coefficients c_k, $k = 1 \ldots r$ are independently distributed random variables according to $u_k(c_k)$, $k = 1 \ldots r$ subject to constraints

$$g_i(x_1 \ldots x_n;\ a_1 \ldots a_q) = 0 \quad (i = 1 \ldots m),$$

where again the a's are independently distributed according to $v_s(a_s)$, $s = 1 \ldots q$. Using Lagrange multipliers, a solution may be obtained of the form

$$x_j = \phi_j(c_1 \ldots c_r;\ a_1 \ldots a_q) \quad (j = 1 \ldots n).$$

The next stage is then to consider a decision function $p(x_1 \ldots x_n)$. The procedure involves taking the expectation $E(p)$ of p with respect to the x_j which are now stochastic variables and maximizing $E(p)$ with respect to parameters $c_1 \ldots c_r;\ a_1 \ldots a_q$ through differentiation. The resulting values of the parameters are the desired estimates. Second derivatives are also used for sufficiency tests.

One of the famous problems of operations research is that of the newspaper boy. We give it here in a slightly different form.

Example 1. The Publisher

A publisher sells a book at a profit of $2.00 a copy and loses $5.00 for each unsold copy. Assuming that the demand is unknown but can be estimated from previous data as being given by the probability density $f(y)$, with a maximum sale of y_0 copies, it is desired to calculate the number of books to be printed in order to maximize the expected profit.

If the publisher prints x copies when the demand is for y copies, then the profit is given by

$$p = \begin{cases} 2y - 5(x-y) = 7y - 5x & \text{if } y < x, \\ 2x & \text{if } y \geqslant x. \end{cases}$$

The expected profit is given by

$$E[p] = \int_0^x (7y - 5x)f(y)\,dy + \int_x^{y_0} 2xf(y)\,dy$$

with

$$\int_0^{y_0} f(y)\,dy = 1.$$

Note that this gives

$$\int_x^{y_0} f(y)\,dy = 1 - \int_0^x f(y)\,dy.$$

Then $E(p)$ is maximized by equating to zero the derivative with respect to x. Thus

$$\frac{dE}{dx} = \frac{d}{dx}\left[\int_0^x (7y - 5x)f(y)\,dy + 2x - 2x\int_0^x f(y)\,dy\right]$$

$$= 7xf(x) - 7xf(x) - 7\int_0^x f(y)\,dy + 2$$

$$= 0$$

or

$$\int_0^x f(y)\,dy = \tfrac{2}{7}.$$

Thus x must be chosen so that the integral has the value $2/7$.

Example 2. Stochastic Linear Programming: Statement of the Problem

If in a linear-programming problem the constraints must be satisfied with some specified probability $\alpha_i (i = 1, 2, \ldots, m)$, the problem becomes: minimize cx subject to

$$P(\textstyle\sum_j a_{ij}x_i \geqslant b_i) \geqslant \alpha_i \quad (i = 1, 2, \ldots, m).$$

Assume that c and the a_{ij} are constraints and that the b_i are random variables. Then the ith constraint may be rewritten as

$$P\left(\frac{\sum a_{ij}x_j - E(b_i)}{\sigma_{b_i}} \geqslant \frac{b_i - E(b_i)}{\sigma_{b_i}}\right) \geqslant \alpha_i,$$

where σ is the standard deviation of b_i.

If we further assume that b_i is normally distributed and let

$$z_i = \frac{b_i - E(b_i)}{\sigma_{b_i}},$$

we have

$$P(z \leqslant K_{\alpha_i}) = \alpha_i$$

where K_{α_i} is determined from

$$\alpha_i = \int_{-\infty}^{K_{\alpha_i}} \frac{1}{\sqrt{2\pi}} e^{-x^2/2} \, dx.$$

This is equivalent to

$$P\left(K_{\alpha_i} \geqslant \frac{b_i - E(b_i)}{\sigma_{b_i}} \right) = \alpha_i.$$

Thus

$$P\left(\frac{\Sigma a_{ij}x_j - E(b_i)}{\sigma_{b_i}} \geqslant \frac{b_i - E(b_i)}{\sigma_{b_i}} \right) \geqslant \alpha_i$$

is true if and only if the following holds:

$$\frac{\Sigma a_{ij}x_j - E(b_i)}{\sigma_{b_i}} \geqslant K_{\alpha_i}$$

or

$$\Sigma a_{ij}x_j \geqslant E(b_i) + K_{\alpha_i}\sigma_{b_i}.$$

The original problem may now be restated as: minimize cx subject to $Ax \geqslant E(b) + K_{\alpha_i}\sigma_{b_i} x \geqslant 0$ which is a deterministic problem, if $E(b)$ and σ_b are known.

4.10.   Game Theory [Luce and Raiffa]

Game theory gives us a method for optimizing the solution of multiple interest problems in which each party pursues its objective or interest while being aware that the others are pursuing these same objectives. The problem is to find a solution which satisfies everyone. (For more on game theory see Chapter 7).

Example. Examples of Utility Computation

(a) In a coin-matching game each player shows one side of his coin, Player I wins one unit if each shows the same side and loses a unit if they do not. The essential point here is the symmetry of the game. We may describe the game by a matrix A where a_{ij} gives the gain to Player I if he follows strategy i while Player II follows strategy j.

		Player II	
		Heads	Tails
Player I	Heads	1	-1
	Tails	-1	1

(b) We may consider a simple example. Two airplanes belonging to Player I wish to destroy a target of Player II to which there are three routes; Player II has three guns that he can place on any of the routes. If an airplane travels over a route on which there is a gun it is put out of action. If two airplanes go over only one gun, only one aircraft is destroyed. Player I wishes to destroy the target and Player II to defend it. Player I has two strategies: (1) he may send each airplane by a different route, or (2) he may send both on the same route. Player II may (1) assign each gun to a different route, (2) assign two guns to one route and one gun to another route, (3) assign all three guns to the same route. We set the payoff for destroying the target at unity, and we let $a_{ij}, i = 1, 2; j = 1, 2, 3$ denote the payoff to I for playing strategy i while II plays strategy j.

We can see that $a_{11} = 0$ and $a_{21} = 1$. We may also show that $a_{12} = 2/3$. Here the guns are allocated to the routes in a series of 0, 1, 2 patterns.

The planes may be assigned to these patterns as follows: $(0, 1, 1)$, $(1, 0, 1)$, and $(1, 1, 0)$, and two of these possibilities use an undefended route. Thus the probability of getting through is 2/3. We also have $a_{22} = 2/3$, since the probability that at least one plane gets through is equal to that of choosing a route not defended by two guns; similarly $a_{23} = 2/3$. We also have $a_{13} = 1$. This gives the payoff matrix

		II	
	A gun on each route	A gun on one route and two on another	Three guns on one route
I Planes on different routes	0	$\frac{2}{3}$	1
Planes on same route	1	$\frac{2}{3}$	$\frac{2}{3}$

In general, it is not easy to write down the payoffs, whether given in cardinal or ordinal form. For comparison of weapons systems designed for different missions, the payoff may be computed by means of simulation in which a weapon system is theoretically tested against the different missions for various measures of effectiveness of destruction, deterrence, reliability, cost, etc. The reader can imagine the added difficulty when the utility scales of the players are different.

The previous examples illustrated zero-sum games in which the gain of Player I has the same magnitude as the loss of Player II. A game is nonzero-sum if the payoffs to the players for a particular choice of strategies are not equal and opposite.

In this chapter we have dealt with a variety of methods for optimization, one of the most common problems we face in our personal affairs as well as in our business and professional life.

Chapter 4—Problems

1. A gardener estimates that if he pulls all his carrots now he will have 300 boxes worth \$5 per box. If he waits, the size of the carrots will increase so as to give him an additional 50 boxes per week, but the price will drop by 50 cents per box per week. When should he dig them? [After two weeks.]

2. The manager of a real estate firm faces the problem of deciding the monthly rental he should charge for each of the 60 newly built apartment units. His past experience tells him that at \$80 rental per month, all apartment units will be occupied but, for each \$2.00 increase in rent, one apartment unit is likely to remain vacant. Since service and maintenance costs are higher for occupied than for vacant apartments, he realizes that he can increase his profit margin by raising the monthly rental from \$80 even though this means that some of the apartments will remain unoccupied. How much rental should he charge to collect the maximum total rent? [\$100]

Now expand your model to take care of maintenance costs. Let A be the maintenance costs for unoccupied apartment and B the maintenance cost for occupied apartment.

3. Find the dimensions of the largest box (sides parallel to the coordinate planes) which may be inscribed in the ellipsoid

$$\frac{x^2}{b^2} + \frac{y^2}{b^2} + \frac{z^2}{c^2} - 1 = 0 \qquad \left(\frac{2a}{3}, \frac{2b}{3}, \frac{2c}{3}\right).$$

4. We wish to satisfy the scheduling requirements of an airline using the smallest number of airplanes. Describe a simple schedule involving nine flights among three cities and formulate a model, (i) using network flow methods, (ii) using linear-programming methods.

5. A rug company is under contract to manufacture a rectangular rug to cover the maximum floor area of an ellipse-shaped hall. The dimensions of the hall will not be known before the completion of the rug, but the probability densities $f(a)$ and $g(b)$ of the axes a, b of the hall are known. The rug company will be paid for the area of the rug that fits the hall with sides parallel to the axes. If the rug is smaller than required, the company will lose in profits. On the other hand, if the rug is larger, it will need cutting to the required dimensions and so will incur a loss due to the unused portion. What size should the rug be?

6. Given n targets distributed over a wide terrain. Each target has any one of three possible values, 1, 2, 3; the greater number corresponds to the higher value. An airplane with fixed gasoline capacity and a maximum payload of 8000 lb wishes to bomb targets on a single mission and return to its base. There are three types of bombs: a 500-lb bomb, which can only destroy targets of value 1; a 1000-lb bomb, which can destroy targets of value 1 or 2; and a 2000-lb bomb, which can destroy targets of value 1, 2, or 3. The plane can carry no more than 12 bombs. The fuel consumption is proportional to the weight of the airplane which depends on the weight of the unexpended bomb load and on the remaining fuel. It is assumed that a bombed target is completely destroyed. The optimal selection of targets determines the order in which bombs are loaded and later dropped. What type of bombs and what order of loading would produce the maximum total value of targets destroyed?

7. Formulate some situations which you have faced in the last month as a two-person game? Were your actions rational in the light of the insights derived from the game. If not, what went wrong?

References

Deakin, Michael A. B., Walking in the rain, *Math Mag.*, Vol. 45, November, 1972.

Jacobs, W. W., The caterer problem, *Naval Res. Logistics Quart.*, Vol. 1, 1954, pp. 154–165.

Koopman, B. O., The theory of search, *Operations Res.*, Vol. 4, No. 3, and Vol. 4, No. 5, 1956.

Luce, R. D. and H. Raiffa, *Games and Decisions*, Wiley, New York, 1957.

Modigliani, F. and F. Holm, Production planning, overtime and the nature of the expectation and planning horizon, *Econometrica*, Vol. 23, 1955, p. 46.

Nash, J., Non-cooperative games, *Ann. Math.*, Vol. 54, No. 2, 1951.

Rosen, Robert, *Optimization Principles in Biology*, Plenum Press, New York, 1967.

Saaty, T. L., *Mathematical Models of Arms Control and Disarmament*, Wiley, New York, 1968.

Saaty, T. L., *Mathematical Methods of Operations Research*, McGraw-Hill, New York, 1959.

Saaty, T. L., *Optimization in Integers and Related Extremal Problems*, McGraw-Hill, New York, 1970.

Saaty, T. L. and J. Bram, *Nonlinear Mathematics*, McGraw-Hill, New York, 1964.

Vidale, M. L. and H. B. Wolfe, An O. R. study of sales response to advertising, *Operations Res.*, Vol. 5, No. 3, June 1957, pp. 370–381.

Bibliography

General

Beveridge, Gordon S. G. and Robert S. Schechter; *Optimization: Theory and Practice*, McGraw-Hill, New York, 1970.
Cooper, Leon, *et al.*, *Introduction to Operations Research Models*, W. B. Saunders Co., Philadelphia, 1977.
Cooper, Leon and David Steinberg, *Introduction to Methods of Optimization*, W. B. Saunders Co., Philadelphia, 1970.
Dantzig, G. B. and B. C. Earnes (eds.), *Studies in Optimization*, (MAA Studies, No. 10), Mathematical Association of America, 1977.
Gillett, Billy E., *Methods of Operations Research*, McGraw-Hill, New York, 1976.
Gupta, Shiv K. and John M. Cozzolino, *Fundamentals of Operations Research for Management*, Holden-Day, San Francisco, California, 1975.
Hancock, Harris, *Theory of Maxima and Minima*, Dover Publications, New York, 1960.
Pierre, Donald A., *Optimization Theory with Applications*, Wiley, New York, 1969.
Saaty, Thomas L., *Mathematical Methods of Operations Research*, McGraw-Hill, New York, 1959.
Shelly, Maynard W. II and Glenn L. Bryan, *Human Judgements and Optimality*, Wiley, New York, 1964.
Wilde, Douglass J. and Charles S. Brightler, *Foundations of Optimization*, Prentice-Hall, Englewood Cliffs, New Jersey, 1967.

Geometric Approach (Convexity)

Bonnesen and Fenchel, *Theorie der Konvexen Korper*, Chelsea Publishing Co., New York, 1948.
Hadwiger, H., *Altes and Neues uber Konvexe Korper*, Basel, Birkhauser, 1955.
Valentine, Frederick A., *Convex Sets*, McGraw-Hill, New York, 1964.
Yaglom and Boltyanskii, *Convex Figures* (translated by P. J. Kelly and L. F. Walton), Holt, Rinehard & Winston, New York, 1961.

Geometric Number Theory

Minkowski, Hermann, *Geometrie Der Zahlen*, Chelsea Publishing Co., New York, 1953.
Rogers, C. S., *Packing and Covering*, Cambridge University Press, London, 1964.
Féjés-Toth, L., *Lagerungen In Der Ebene Auf Der Kugel Und Im Raum*, Springer-Verlag, Berlin–Gottingen–Heidelberg, 1953.

Geodesics, Differential Geometry and Manifolds

Flanders, Harley, *Differential Forms, with Applications to the Physical Sciences*, Academic Press, New York, 1963.
Graustein, William C., *Differential Geometry*, Macmillan, New York, 1949.

Network Flow

Berge, Claude, *The Theory of Graphs*, Methuen, London; Wiley, New York, 1962.
Berge, C. and A. Ghouila-Houri, *Programmes, Jeux Et Reseaux De Transport*, Dunod, Paris, 1962.
Busacker, Robert G. and Thomas L. Saaty, *Finite Graphs and Networks: An Introduction with Applications*, McGraw-Hill, New York, 1965.
Ford, L. R., Jr., and D. R. Fulkerson, *Flow in Networks*, Princeton University Press, Princeton, New Jersey, 1962.

Equality and Inequality Constraints

Beckenbach, Edwin and Richard Bellman, *Inequalities*, Springer-Verlag, Berlin, 1961.
Hardy, C. H., J. E. Littlewood, and G. Polya, *Inequalities*, Cambridge University Press, 1934.
Kazarinoff, N. D., *Analytic Inequalities*, Holt, Rinehart & Winston, New York, 1961.
Kuhn, H. W. and A. W. Tucker, *Linear Inequalities and Related Systems*, No. 38, Princeton University Press, Princeton, New Jersey, 1956.

Saaty, Thomas L., *Modern Nonlinear Equations*, McGraw-Hill, New York, 1964.

References Containing a Short Account of the Calculus of Variations (Usually with Applications)

Bliss, G. A., *A Monograph on the Calculus of Variations*, Open Court Publishing Co., Chicago, 1935. MAA Carus monograph.
Byerly, W. E., *Introduction to the Calculus of Variations*, Ginn & Co., Boston, Massachusetts, 1917.
Detlman, J. W., *Mathematical Methods in Physics and Engineering* (ch. 2), McGraw-Hill, New York, 1962.
Hildebrand, F. B., *Methods of Applied Mathematics* (ch. 2), Prentice-Hall, Englewood Cliffs, New Jersey, 1952.
Hestenes, M. R., Elements of the calculus of variations, pp. 59–91 in *Modern Mathematics for the Engineer*, Vol. I. E. F. Bekenback (ed.), McGraw-Hill, New York, 1956 (reprinted in McGraw-Hill paperbacks).
Sagan, H., *Boundary and Eigenvalue Problems in Mathematical Physics*, Wiley, New York, 1951.

References with Emphasis on the Applications of the Calculus of Variations to Problems in Geometry, Mechanics, Partial Differential Equations and Mathematical Physics

Courant, R., *Dirichlet Principle. Conformal Mapping and Minimal Surfaces*, Interscience, New York, 1950.
Goldstein, Herbert, *Classical Mechanics*, Addison-Wesley, Reading, Massachusetts, 1950.
Lanczos, C., *The Variational Principles of Mechanics*, University of Toronto Press, Toronto, 1949.

Applications to Dynamic Programming, the Maximum Principle of Pontryagin, Control and Optimization Theory and Related Numerical Algorithms

Athan, Michael and Peter Falb, *Optimal Control*, McGraw-Hill, New York, 1966.
Balakrishnan, A. V. and L. W. Neustadt (eds.), *Computing Methods in Optimization Problems*, Academic Press, New York, 1964.
Bellman, R., *Dynamic Programming*, Princeton University Press, Princeton, New Jersey, 1961.
Bellman, R., *Adaptive Control Processes: A Guided Tour*, Princeton University Press, Princeton, New Jersey, 1961.
Bellman, R. and Dreyfus, S. E., *Applied Dynamic Programming*, Princeton University Press, Princeton, New Jersey, 1962.
Bellman, R. E. (ed.), *Symposium on Mathematical Optimization Techniques*, the RAND Corp., 1963.
Bellman, R. E. and Kalaba, R. (eds.), *Selected Papers on Mathematical Trends in Control Theory*, Dover, New York, 1964.
Bellman, R. E., L. Glicksberg, and O. A. Gross, *Some Aspects of the Mathematical Theory of Control Processes*, the RAND Corp.
Feldbaum, A. A., *Optimal Control Systems*, Academic Press, New York, 1966.
Advances in Control Systems, a series edited by C. T. Leondes, Academic Press, New York, 1964.
Leitmann, G., *Optimization of Techniques with Applications to Aerospace Systems*, Academic Press, New York, 1962.
Merriam, C. W., III., *Optimization Theory and the Design of Feedback Control Systems*, McGraw-Hill, New York, 1964.
Pontryagin, L. S., *The Mathematical Theory of Optimal Processes*, (translated from the Russian), Interscience, New York, 1962.
Saaty, T. L. and J. Bram, *Nonlinear Mathematics*, McGraw-Hill, New York, 1964.
Tou, J. L., *Modern Control Theory*, McGraw-Hill, New York, 1964.
Wilde, D. J., *Optimum Seeking Methods*, Prentice-Hall, Englewood Cliffs, New Jersey, 1964.

Morse Theory and the Calculus of Variations in the Large

Morse, M., *Calculus of Variations in the Large*, American Math. Society Colloquium Publications, Vol. 18, Providence, Rhode Island, 1934.
Liusternik, L. A., *Topologies of Functional Spaces and Calculus of Variations in the Large*, Moscow, 1947.
Seifert, H. and W. Threlfall, *Variations rechung in Grossen* (Theorie von Marston Morse), Chelsea Publication Co., New York, 1951.

Functional Analysis and Modern Variational Theory

Liusternik, L. and Sobolev, V., *Elements of Functional Analysis* (ch. 6) (translated from the Russian), Ungar Publishing Co., New York, 1961.
Mikhlin, S. G., *The Problem of the Minimum of a Quadratic Functional* (translated from the Russian by A. Feinstein), Holden-Day, San Francisco, California, 1965.
Mikhlin, S. G., *Variational Methods in Mathematical Physics* (translated by L. I. Chambers), Pergamon Press, Oxford, 1964.
Vainberg, M. M., *Variational Methods for the Study of Nonlinear Operators* (translated by Amiel Feinstein), Holden-Day, San Francisco, California, 1964.
Kantorovich, L. V., and Akilov, G. P., *Functional Analysis in Normed Spaces*, Moscow, 1959. English translation edited by A. P. Robertson, Pergamon Press, Oxford, 1964.

Expository papers and Symposia

Bliss, G. A., Some recent developments in the calculus of variations, *Bull. Am. Math. Soc.*, Vol. 26, 1920, pp. 343–361.
Dresden, A., Some recent work in the calculus of variations, *Bull. Am. Math. Soc.* Vol. 30, 1926, pp. 475–521.
University of Chicago Department of Mathematics, *Contributions to the Calculus of Variations*, University of Chicago Press, Chicago, 1930–1932.

Some References to More Recent Research on the Calculus of Variations and Its Application

Proceedings of the Eighth Symposium in Applied Mathematics of The American Math Society, Vol. VIII, *Calculs of Variations and Its Applications*, McGraw-Hill, New York, 1958.
Morrey, C. B., Jr., *Multiple Integral Problems in the Calculus of Variations and Related Topics*, University of California Press, Berkeley and Los Angeles, 1943.
Morrey, C. B., Jr., *Multiple Integrals in the Calculus of Variations*, Am. Math. Soc. Colloquium Lectures, August 25–28, 1964. (Contains an extensive bibliography on books and recent research papers.)
Rothe, E. H., Gradient mappings, *Bull. Am. Math. Soc.*, Vol. 59, 1953, pp. 5–19.
Rothe, E. H., Remarks on the applications of gradient mappings to the calculus of variations and connected boundary value problems in partial differential equations, *Common Pure and Applied Math.*, Vol. 9, 1956, pp. 551–568.

Linear and Nonlinear Programming

Arrow, Kenneth J., L. Hurwicz, and H. Uzawa, *Studies in Linear and Nonlinear Programming*, Stanford University Press, Stanford, California, 1958.
Charnes, A. and W. W. Cooper, *Management Models and Industrial Applications of Linear Programming*, Vols. I and II, Wiley, New York, 1961.
Dantzig, G. B., *Linear Programming and Extensions*, Princeton University Press, Princeton, New Jersey, 1964
Ficken, F. A., *The Simplex Method of Linear Programming*, Holt, Rinehart & Winston, New York, 1961.
Gale, D., *The Theory of Linear Economic Models*, McGraw-Hill, New York, 1960.
Gass, S. I., *Linear Programming*, 2nd edn., McGraw-Hill, New York, 1964.
Hadley, G., *Nonlinear and Dynamic Programming*, Addison-Wesley, Reading, Massachusetts, 1964.
Hadley, G., *Linear Programming*, Addison-Wesley, Reading, Massachusetts, 1962.
Kunzi, Hans P. and Wilhelm Kreller, *Nichtlineare Programmieung*, Springer-Verlag, Berlin–Gottingen–Heidelberg, 1962.
Riley, Vera and S. I. Gass, *Linear Programming and Associated Techniques*, Operations Research Office, Johns Hopkins University, Chevy Chase, Maryland, 1958.
Simonnard, M., *Programmation Linearirer*, Dunod, Paris, 1962.
Wilde, Douglass J., *Optimum Seeking Methods*, Prentice-Hall, Englewood Cliffs, New Jersey, 1964.
Wolfe P. and R. Graves, *Recent Advances in Mathematical Programming*, McGraw-Hill, New York, 1963.
Zoutendijk, G., *Methods of Feasible Directions*, Elsevier, Amsterdam, 1960.

Dynamic Programming

Aris, Rutherford, *Discrete Dynamic Programming*, Blaisdell Publishing Co., New York, 1964.
Dreyfus, S. E., *Dynamic Programming and the Calculus of Variations*, Academic Press, New York, 1965.
Kaufmann, A. and R. Cruon, *La Programmation Dynamique*, Dunod, Paris, 1965.

Discrete Approach to Problems

Cassels, J. W. S., *Diophantine Approximation*, Cambridge University Press, 1965.
Ryser, Herbert J., *Combinatorial Mathematics*, The Mathematics Association of America, 1963.
Saaty, Thomas L., *Optimization in Integers*, McGraw-Hill, New York, 1970.

Game, Decision, and Value Theories

Debreu, Gerard, *Theory of Value*, Wiley, New York, 1959.
Dresher, Melvin, *Games of Strategy Theory and Applications*, Prentice-Hall, Englewood Cliffs, New Jersey, 1961.
Dresher, M., A. W. Tucker, and P. Wolfe, *Contributions to the Theory of Games*, Vol. III, Princeton University Press, Princeton, New Jersey, 1957.
Harsanyi, J. C., *Rational Behavior and Bargaining Equilibrium in Games and Social Situations*, Cambridge University Press, 1977.
Isaacs, Rufus, *Differential Games*, Wiley, New York, 1965.
Karlin, Samuel, *Mathematical Methods and Theory in Games, Programming, and Economics*, Vols. I and II, Addison-Wesley, Reading, Mass., 1959.
Kuhn, H. W. and A. W. Tucker, *Contributions to the Theory Games*, Vol. I, Princeton University Press, Princeton, New Jersey, 1950.
Luce, R. Duncan and Howard Raiffa, *Games and Decisions*, Wiley, New York, 1957.
Owen, Guillermo, *Game Theory*, W. B. Saunders Co., Philadelphia 1968.
Rapoport, Anatol, *Two-person Game Theory, The Essential Ideas*, University of Michigan Press, 1966.
Rapoport, *N-Person Game Theory Concepts and Applications*, University of Michigan Press, 1970.
Shubik, Martin, *Strategy and Market Structure*, Wiley, New York, 1959.
Tucker, A. W. and R. D. Luce, *Contributions to the Theory of Games*, Vol. IV, No. 40, Princeton University Press, Princeton, New Jersey, 1959.
Vajda, S., *Mathematical Programming*, Addison-Wesley, Reading, Massachusetts, 1961.
Vajda, S., *The Theory of Games and Linear Programming*, Methuen, London and Wiley, New York, 1956.
Ventzel, E. S., *Lectures on Game Theory*, Vol. VI, Gordon & Breach, New York, 1961.
von Neumann, John and Oskar Morgenstern, *Theory of Games and Economic Behavior*, Princeton University Press, Princeton, New Jersey, 1944.

Chapter 5

Probability and Stochastic Processes

5.1. Introduction

HERE we pass from deterministic models of optimization to models which use techniques from the theory of probability and from the quantitative study of inductive inference. By assigning a number, a probability, between zero and unity to the occurrence of an event, one has a measure of the likelihood of its occurrence. For example, unity indicates that the event will occur with certainty.

Probability theory provides a variety of useful models suitable for the description and interpretation of complex problems. As in other applications of mathematics, there is usually sufficient closeness between the ideal and the real to make the results useful. Repetition of a process with varying outcomes enables the detection of order, the formulation of a model, and subsequently the development of a theory with selected measures of effectiveness. This representation should be and usually is checked against actual data. When success, within prescribed limits, has been attained, the theory may then be used for prediction purposes.

In any case an application of the theory of probability generally derives its concepts from the underlying physical situation and proceeds to develop, as any nonabstract correct theory does, a logically sound system from which reliable answers may be deduced.

In our models, we shall assume that the reader is familiar with the basic concepts of probability.

The physical situations which prompted the development of the theory of stochastic processes are numerous. Time-ordered sequences of related random events occur commonly. The stock market averages for the coming days may be considered as random variables. If we wish to project several days ahead, we must determine the joint distribution of the stock market averages for those days, knowing the past history. Situations arise in which the relationship among random events depends both on time and on space: a simple example is that of causing a ripple in a still pond. The water level at any fixed point depends both on the time after the initial disturbance and the position of the point. The general model, a set of well-defined random variables with a known joint distribution, is again a useful construct.

While knowledge of the set of random variables and their joint distribution (or the joint distributions of all finite subsets if the number of random variables is infinite) is equivalent to complete information about any reasonable process, it is often not practical or possible to derive the joint distributions mathematically. To obtain an adequate resolution frequently requires much less information. Thus, a solution to a stochastic process does not always mean total knowledge of the joint distributions. Much of the work in stochastic processes has been directed towards the derivation of characteristics of a process without resort to the joint distributions.

In preparation for understanding the general ideas, probability measures and stochastic events need to be defined. The starting point is the basic space U called the sample space. The elements of U are the possible "occurrences," "elementary events," "outcomes," or "effects." Associated with U is a collection of measurable subsets B of U. The collection B forms a special kind of set called a sigma-field. A measure P, called a probability measure, is defined on the elements of B. A real-valued function x from U to the real numbers R, measurable with respect to B, is called a random variable. x is B measurable if and only if $\{\omega : x(\omega) \leqslant a\} \in B$ for all $a \in R$. Associated with each random variable is a distribution function F, where

$$F(a) = P\{x(\omega) \leqslant a\}; \quad a \in R; \quad \omega \in U.$$

$F(a)$ is well-defined since the set, $\{x(\omega) \leqslant a\} \in B$ by the measurability of x. The introduction of a random variable and its distribution function allows us to translate the actual physical events into numerical form, in much the same way as a functional equation gives an explicit algebraic form for a cause–effect relationship. In fact, once the random variable is well defined, the underlying sigma-field and probability measure may be dropped. The range of the random variable will replace the set of physical events as the sample space, and all mathematical development will be performed at this level.

The joint distribution of a family of random variables $(x_1, x_2, \ldots, x_n)$ is a multivariate function defined as follows:

$$F(a_1, a_2, \ldots, a_n) = P\{x_1(\omega) \leqslant a_1, \quad x_2(\omega) \leqslant a_2 \ldots x_n(\omega) \leqslant a_n\};$$
$$a_i \in R; \quad \omega \in U; \quad \{x_1(\omega) \leqslant a_1, \quad x_2(\omega) \leqslant a_2 \ldots x_n(\omega) \leqslant a_n\} \in B.$$

In order to characterize stochastic processes, we start with a family of random variables. The elements which distinguish the various stochastic processes are the state space, the index set, and the set of all finite dimensional distribution functions. The state space is the subset of Euclidean n-space over which the values of the random variable range; the index set merely indexes and, in some cases, orders the random variables; the distributions give the relations between the random variables. These three elements are used to classify all stochastic processes.

We shall characterize four major types of stochastic processes:

(i) *Independent increments.* Given any finite subset of the index set, say $(t_1, t_2, \ldots, t_n)$, such that $t_1 < t_2 < \ldots < t_n$, then if the random variables $x_{t_2} - x_{t_1}$, $x_{t_3} - x_{t_2}, \ldots, x_{t_n} - x_{t_{n-1}}$ are independent, the result is a stochastic process with independent increments. In such a process, the joint distributions of any finite set of x_{t_i}'s can be found easily, for $x_{t_i} = z_{t_1} + z_{t_2} + \ldots + z_{t_i}$, where $z_{t_i} = x_{t_i} - x_{t_{i-1}}$, and the z_{t_i}'s are independent.

(ii) *Martingales.* Given any finite subset of the index set, say $(t_1, t_2, \ldots, t_n, t_{n+1})$, such that $t_1 < t_2 < \ldots < t_n < t_{n+1}$; then if $E(x_{t_{n+1}} | x_{t_1} = a_1, x_{t_2} = a_2 \ldots x_{t_n} = a_n) = a_n$, the stochastic process is called a martingale. (This is the basic condition for a fair game.)

(iii) *Markov processes.* Given any subset of the index set, say $(t_1, t_2, \ldots, t_n)$, such that $t_1 < t_2 < \ldots < t_n$, then if $P(x_{t_n} \leqslant a_n | x_{t_1} = a_1, x_{t_2} = a_2 \ldots x_{t_{n-1}} = a_{n-1}) = P(x_{t_n} \leqslant a_n | x_{t_{n-1}} = a_{n-1})$, the stochastic process is called a Markov process.

(iv) *Stationary processes.* Given any finite subset of the index set, say

$(t_1, t_2, \ldots, t_n)$, and any $h > 0$, then if the joint distributions of $x_{t_1}, x_{t_2}, \ldots, x_{t_n}$, and $x_{t_1} + h,\ x_{t_2} + h, \ldots, x_{t_n} + h$ are the same, the stochastic process is a stationary process.

In practice, the existence of some stable limiting distribution may be important to us. The occurrence or recurrence of a particular state may be of interest. Sometimes one may wish to find the expected first passage times or recurrence times for given states. For specialized processes, existence theorems and methods of analysis exist for determining these characteristics without the necessity of deriving the complete joint distribution.

The use of stochastic processes lends itself easily to the following major areas of application: (i) queueing theory, (ii) inventory theory, (iii) reliability theory, (iv) renewal theory, and (v) stochastic optimization.

5.2. Applications of the Theory of Probability

We now give some elementary examples which make use of probability concepts. Our first two examples deal with very simple circuit and switching problems.

Example 1. Circuits and Switches

In the circuit in Fig. 5.1, what is the probability that the bulb will be lit (i.e., the circuit closed), given that it is equally likely for any of the switches A, B, C, D to be open or closed.

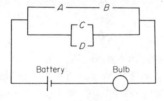

FIG. 5.1.

The bulb will be lit if both switches A and B are closed or if either switch C or switch D is closed. Thus the desired probability is given by

$$P(\{A \text{ and } B\} \text{ or } C \text{ or } D) = P(AB) + P(C) + P(D)$$
$$- P(ABC) - P(ABD) - P(CD) + P(ABCD) = 13/16.$$

The probability of any switch being closed is $1/2$; and thus, $P(C) = P(D) = 1/2$. We also have, $P(AB) = P(CD) = 1/4$, $P(ABC) = P(ABD) = 1/8$, and $P(ABCD) = 1/16$ since, because of independence, the probabilities are given by the products of the corresponding single-switch probabilities.

Example 2. Craps

In this game the "passer" rolls out two dice and wins if he makes 7 or 11 on the first roll (these are called "naturals"). He loses if he makes 2, 3, or 12. If he makes 4, 5, 6, 8, 9, 10 on the first roll, then he rolls until he makes the same point or gets 7. If he rolls 7 before getting the same point, he loses; otherwise, he wins. What is the probability of winning?

First, consider the probability of obtaining a particular sum on a single roll of two dice. The probability of turning up a particular point on each dice is clearly 1/6. Then the probability of obtaining a 2 is $1/6 \times 1/6 = 1/36$, by the compound-probability property of independent events. Now for a 3, one can have a 1 and 2 or 2 and 1. The probability is then $1/36 + 1/36 = 1/18$, applying the properties of compound and total probability. For 4, one can have a 1 and 3, 3 and 1, 2 and 2 for a probability of 1/12. By considering all possible sums in this fashion, one has the probabilities given as follows:

No.	Probability	No.	Probability
2	1/36	7	1/6
3	1/18	8	5/36
4	1/12	9	1/9
5	1/9	10	1/12
6	5/36	11	1/18
		12	1/36

The passer wins if he obtains one of the following:

1. A 7 or an 11 on the first roll, with probability P_A.
2. A repetition of any other number (a 4,5,6,8,9,10) appearing on the first roll, before a 7 is rolled. Let this probability be P_B. Now

$$P_{\text{win}} = P_A + P_B,$$

where

$$P_A = \frac{6}{36} + \frac{2}{36} = \frac{8}{36} = \frac{2}{9}.$$

To calculate P_B, note that, having obtained a 4 on the first roll, the probability of obtaining another 4 on a subsequent roll before casting a 7 is the sum of the probabilities of obtaining a 4 on the first roll and on the second, obtaining a 4 on the first roll, not obtaining a 4 or a 7 on the second, then obtaining a 4 on the third roll, etc. This gives

$$\frac{3}{36} \times \frac{3}{36} + \frac{3}{36}\left(1 - \frac{3}{36} - \frac{6}{36}\right)\frac{3}{36} + \frac{3}{36}\left(1 - \frac{3}{36} - \frac{6}{36}\right)^2 \frac{3}{36} + \cdots$$

Simplification gives

$$\left(\frac{3}{36}\right)^2 \left(1 + \frac{3}{4} + \left(\frac{3}{4}\right)^2 + \cdots\right) = \frac{1}{36}.$$

An alternative approach is as follows. There are nine ways of obtaining 4 or 7, three of which give 4; since the probability of the latter is 3/36, one has

$$\frac{3}{36} \times \frac{3}{9} = \frac{1}{36}.$$

This argument may be applied to the other numbers: the probability of winning on 5 is

$$\frac{4}{36} \times \frac{4}{10} = \frac{8}{5 \times 36};$$

the probability of winning on 6 is

$$\frac{5}{36} \times \frac{5}{11} = \frac{25}{11 \times 36};$$

the probability of winning on 8 is the same as on 6; the probability of winning on 9 is the same as on 5; the probability of winning on 10 is the same as on 4.

$$P_B = 2\left(\frac{1}{36} + \frac{8}{5 \times 36} + \frac{25}{11 \times 36}\right) = \frac{2}{36}\left(1 + \frac{8}{5} + \frac{25}{11}\right) = \frac{1}{18} \times \frac{268}{55}$$

$$P_{win} = \frac{2}{9} + \frac{1}{18} \times \frac{268}{55} = \frac{220 + 268}{990} = \frac{244}{495}.$$

Thus the probability of winning at craps is less than one-half.

Example 3. What is Random?—The Subway Paradox

The Red Line Subway in Boston runs from Harvard Square to Quincy Center via Park Street (Fig. 5.2). Each day you arrive at the Park Street Station at a random time, go down

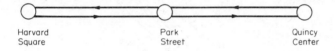

Harvard   Park   Quincy
Square   Street   Center

Fɪɢ. 5.2.

to the central platform, and get on the first Red Line train to arrive. Nine days out of ten you end up at Harvard Square. How can this be?

Most people's intuition tells them that one should go to Quincy Center as often as one goes to Harvard Square. This is not necessarily the case. The error results from considering the wrong property as being random.

Insight: notice that it is you, and not the trains, which arrives at random. Imagine, for example, that trains from Harvard Square to Quincy Center run every 10 minutes on a regular schedule. Also, to avoid having a build-up of trains at one end of the line or the other, trains also run from Quincy Center to Harvard Square every 10 minutes.

Solution: to resolve the paradox, imagine that each train bound for Quincy Center arrives 1 minute after the train bound for Harvard Square.

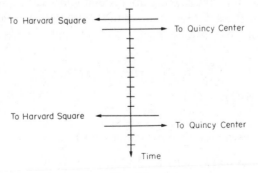

Fɪɢ. 5.3.

Clearly, if your random arrival time falls anywhere between the time of a Quincy Center bound train and a Harvard Square bound train (9 minutes) you will end up at Harvard Square; while if your arrival falls between the time of a Harvard Square bound train and a Quincy Center bound train (1 minute), you will end up at Quincy Center (Fig. 5.3).

Example 4. Meeting of Friends

Two friends, *X* and *Y*, plan to meet downtown between 12:00 noon and 1:00 p.m. Their arrival times are random variables. They agree that each will wait 10 minutes, and that if the other has not arrived during these 10 minutes he will leave. What is the probability that they will meet? Assume that if either arrives after 12:50 he will leave at 1:00 even though he has not waited the full 10 minutes.

We may show the situation graphically as in Fig. 5.4.

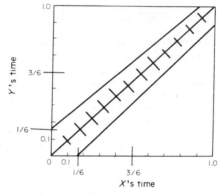

Fɪɢ. 5.4.

Let *x* be the arrival time for *X*, $0 \leqslant x \leqslant 1$. Let *y* be the arrival time for *Y*, $0 \leqslant y \leqslant 1$. Then $y = x$ represents the event that both arrive together.

$$y = x + 1/6, \quad 0 \leqslant x \leqslant 5/6,$$

gives a bound to the situation that *x* arrives first and waits 10 minutes.

$$y = x - 1/6, \quad 1/6 \leqslant x \leqslant 1,$$

gives a bound to the situation where *y* arrives first and waits 10 minutes.

The region between these two lines represents the situation in which the two meet. Since they arrive at random over the hour, the probability is represented by the area of this belt, which is

$$1 - 2(1/2)(5/6)^2 = 11/36.$$

We now consider a risk problem.

Example 5. Risk of Flying on Two-engine and on Four-engine Planes

Is it after to fly on a plane which has two engines or on a plane with four engines?

Let *q* be the probability of failure of a single engine and let $p = 1 - q$ be the probability of successful operation of a single engine. We assume that a four-engine plane can fly on four or on three engines but that a two-engine plane needs both engines. Assuming independence, the probability of four successful engines is p^4, and the probability of exactly three successful engines is $4p^3(1-p)$. For the other plane, the probability of two successful engines is p^2. We wish to determine the range of values of p such that $p^4 + 4p^3 (1-p) \geqslant p^2$; i.e., we seek the range of values of p for which the four-engine plane would be preferred to the two-engine plane. The condition is equivalent to

$$p^2(4p - 3p^2 - 1) \geqslant 0 \quad \text{or}$$
$$3p^2 - 4p + 1 \leqslant 0,$$
$$(3p - 1)(p - 1) \leqslant 0.$$

Since $p \leqslant 1$ we have $3p \geqslant 1$ or $p \geqslant 1/3$. Thus the four-engine plane is safer if $1/3 \leqslant p \leqslant 1$ and the two-engine plane is safer if $0 \leqslant p \leqslant 1/3$. (At $p = 1/3$, each is equally safe.)

These results seem contrary to intuition because the higher the probability of successful operation of a single engine, the more one should ride the four-engine plane; the smaller probability, the more one should ride the two-engine plane. Intuitively it would seem safer to do the opposite.

One must remember that this result depends on our assumptions about safety. The problem might be modified by considering the four-engine plane as safe if at least one engine on each side is operating successfully; in this case it turns out that the four-engine plane is always safer.

Example 6. Bayes' Theorem—Causes from effects or *a priori* from *a posteriori*

For our next example we use Bayes' theorem, which obtains the probability of causes, given the probability of the effects. It often happens that one may wish to isolate the most likely cause to have given rise to an effect when all the possible causes are known and the conditional probability that if a certain cause occurs, then the effect is also known. This probability is determined empirically.

Suppose that an event (a cause) B consists of finite or countably infinite independent events B_i. (The B_i may be regarded as the different ways in which an event can occur.) Then $P(B) = \Sigma P(B_i) = 1$. Let A be an event (an effect) and let $P(B)$ be the *a priori* probability of occurrence of A and $P(A|B_i)$ be the conditional probability of its occurrence, given that B_i has occurred; then the conditional probability of the cause B_i, given that A has occurred, is obtained as follows:

$$P(B_iA) = P(B_i)P(A|B_i) = P(A)P(B_i|A).$$

Thus the last two expressions give

$$P(B_i|A) = \frac{P(B_i)P(A|B_i)}{P(A)}.$$

But
$$P(A) = P(BA) = P(\Sigma B_iA) = \Sigma P(B_iA) = \Sigma P(B_i)P(A|B_i).$$

Thus the probability of the prior occurrence of cause B_i, given that the effect A has occurred, is

$$P(B_i|A) = \frac{P(B_i)P(A|B_i)}{\Sigma P(B_i)P(A|B_i)},$$

which is known as Bayes' theorem.

Example. A ball was transferred from an urn containing two white and two black balls to another containing three white and two black balls. A white ball was then drawn from the second urn. What is the probability that the transferred ball was white?

Let B_1 and B_2 be the two events of transferring a white ball and a black ball, respectively, and let A be the event of drawing a white ball from the second urn.

Now
$$P(B_1|A) = \frac{P(B_1)P(A|B_1)}{\sum\limits_{i=1}^{2} P(B_i)P(A|B_i)},$$

$$P(B_1) = 1/2, \quad P(A|B_1) = 2/3, \quad P(B_2) = 1/2, \quad P(A|B_2) = 1/2.$$

Thus
$$P(B_1|A) = \frac{1/2 \times 2/3}{(1/2 \times 2/3) + (1/2 \times 1/2)} = \frac{4}{7}.$$

5.3. Applications of the Theory of Stochastic Processes

We now give some elementary examples which make use of the theory of stochastic processes. Our first example is better known as Random Walk. We give it here as relating to a drunkard, but it has also been applied to a young man killing time while waiting for his date to show up.

Example 1. The Drunkard

Suppose we find a drunken old man wandering his way up and down the street after leaving his favorite tavern. The street runs east–west. His possible transitions would include taking a step in an easterly direction, taking a step in a westerly direction, or falling flat on his face, and thus not going in any direction. Our street is illustrated in Fig. 5.5. The tavern is at the center.

West _____ Tavern _____ East

FIG. 5.5.

Now we wish to determine the probability of his being at a position k steps away after s transitions. If we assume his probability of movement is the same regardless of his position, we may let p be the probability of an eastward step and q and r be the probabilities of a westward step and of falling down, respectively. Let $P_{k,s}$ be the probability of being at position k after s transitions. There are only three ways we can get there in the sth transition, as shown in Table 5.1.

TABLE 5.1

Position after $(s-1)$ transitions	Probability of given location	Probability of movement to k	Desired probability
$k-1$	$P_{k-1,s-1}$	p	$pP_{k-1,s-1}$
k	$P_{k,s-1}$	r	$rP_{k,s-1}$
$k+1$	$P_{k+1,s-1}$	q	$qP_{k+1,s-1}$
Two or more steps away	$P_{k+2,s-1}$	0	0

Thus we have the relationship

$$P_{k,s} = pP_{k-1,s-1} + rP_{k,s-1} + qP_{k+1,s-1}.$$

We now give a very general example which has many specific uses.

Example 2. Point Processes

Consider a system which can be in any one of n states $E_1 \ldots E_n$. Chance may enter the system in two ways: first, by fixing the times at which the system changes its states (such changes are called transitions), and, second, by fixing the order of the transition among the states. The probabilistic selection of instants of time at which the transitions take place suggests the use of the name "process" and because of the discrete instants involved it is called a point process.

Suppose that the transition events are independent of each other and of the time at which they occur; i.e., the probability of a transition during an interval $(t, t+h)$ does not depend on other transitions (and thus does not depend on what occurred before the instant t) nor on the time t. The generation of telephone calls in a switch board and the arrivals of ships in a harbor satisfy this assumption. Thus, the probability that no event (transition) occurs during the time interval $(t, t+h)$ depends only on h. Denote this probability by $p(h)$. The probability that no event occurs during the time interval $(t, t+h+k)$ is then $p(h+k)$. The probability that no event occurs in $(t, t+h)$ followed by no event occurring in $(t+h, t+h+k)$ is $p(h)p(k)$, by our assumption that these two events are independent. Thus, for any h and k we have the functional equation describing the probability of nonoccurrence of transitions

$$p(h+k) = p(h)p(k).$$

The solution of this functional equation is given by $p(t) = e^{at}$. Since $0 \leqslant p(t) \leqslant 1$, we must have $a = -\lambda$, with λ positive. If t_i and t_{i+1} are the successive instants of transition to states E_i and E_{i+1}, respectively, then

$$\text{prob}\,(t_{i+1} - t_i < t) = 1 - e^{-\lambda t}.$$

For n intervals separating $n+1$ consecutive transition times, we obtain the probability as follows:

Let
$$x = \lambda(t_{i+1} - t_i),$$
$$y = \lambda(t_{i+2} - t_{i+1}),$$
so that
$$x + y = \lambda(t_{i+2} - t_i).$$

Then
$$f_1(x)\,dx = e^{-x}\,dx,$$
$$f_1(y)\,dy = e^{-y}\,dy,$$

and for
$$u = x + y$$
we have

$$f_2(u)\,du = \int_{x=0}^{u} (e^{-x}\,dx)(e^{-(u-x)}\,du) = ue^{-u}\,du.$$

Extending this to n successive time intervals, where

$$u = \lambda(t_{i+n} - t_i)$$

yields

$$f_n(u)\,du = e^{-u}\frac{u^{n-1}}{(n-1)!}du \quad (u > 0),$$

which may be proved by induction.

Now let $t = t_{i+n} - t_i$, so that $u = \lambda t$. Then

$$e^{-u}\frac{u^{n-1}}{(n-1)!}du = e^{-\lambda t}\frac{(\lambda t)^{n-1}}{(n-1)!}\lambda\,dt \quad (t > 0).$$

Now if p_n = probability of exactly n events in time T and π_n = probability of at least n events in time T, then

$$\pi_n = \int_0^T e^{-\lambda t}\frac{(\lambda t)^{n-1}}{(n-1)!}\lambda\,dt.$$

Since $p_n = \pi_n - \pi_{n+1}$, we obtain

$$p_n = \frac{e^{-\lambda T}(\lambda T)^n}{n!} \quad (n \geq 0),$$

which is the Poisson process with parameter λT.

There are more general definitions of point processes than the definition which we have given, but this is adequate for our purpose here.

Note that a Poisson process is an example of a continuous time ($T = [0, \infty]$) stochastic process and is a special case of a Markov process which we discuss next. The sample function x_t counts the number of times a specified event occurs during the time period from 0 to t.

Concrete examples of such processes are the number of X-rays emitted by a substance undergoing radioactive decay, the number of telephone calls originating in a given locality, the occurrences of accidents at a given intersection, the occurrence of errors in a page of typing, breakdowns of a machine, and the arrival of customers for service. The justification for viewing these examples as Poisson processes is based on the concept of the law of rare events. We have a situation with several Bernoulli trials, each with a small probability of success, where the expected number of successes is constant. Under these conditions the actual number of events follows a Poisson law.

The Poisson process has the following characteristics:

1. Suppose $t_0 < t_1 \ldots < t_n$. The increments $X_{t_1} - X_{t_0}, \ldots, X_{t_n} - X_{t_{n-1}}$ are mutually independent random variables.
2. The probability distribution of $X_{t_2} - X_{t_1}$, $t_2 > t_1$, depends only on $t_2 - t_1$, and not, for example, on t_1.
3. The probability of at least one event occurring in an interval of time t is given by $\lambda t + o(t)$.

A more general derivation of the Poisson process (not necessarily homogeneous) may be obtained by assuming that the transitions in fact depend on the state, i.e., the probability of a transition in the interval $(t, t + \Delta t)$ from state E_n to state E_{n+1} is $\lambda_n \Delta t + o(\Delta t)$.

Let $P_n(t)$ be the probability of being in E_n at time t. Then

$$P_n(t + \Delta t) = P_n(t)(1 - \lambda_n \Delta t) + P_{n-1}(t)\lambda_{n-1} \Delta t + o(\Delta t).$$

Then

$$P_n(t + \Delta t) - P_n(t) = -\lambda_n \Delta t \, P_n(t) + \lambda_{n-1} \Delta t \, P_{n-1}(t) + o(\Delta t)$$

and

$$\lim_{\Delta t \to 0} \frac{P_n(t + \Delta t) - P_n(t)}{\Delta t} = -\lambda_n P_n(t) + \lambda_{n-1} P_{n-1}(t).$$

From the above expression, we have

$$P_n'(t) = -\lambda_n P_n(t) + \lambda_{n-1} P_{n-1}(t) \quad (n \geqslant 1)$$

and

$$P_0'(t) = -\lambda P_0(t), \quad P_0(0) = 1.$$

We may now let $\lambda_n = \lambda$ for all n; i.e., the probability of change is assumed constant regardless of the state. The solution of this system again gives the Poisson law

$$P_n(t) = \frac{(\lambda t)^n}{n!} e^{-\lambda t}.$$

We now consider another general example with many specific uses.

Example 3. **Markov Processes and Markov Chains; Semi-Markov Processes**

Suppose we have a system which can be in any of n possible states $E_1 \ldots E_n$. We are interested in state changes which occur in time according to some probability law. Denote the state of the process at time t by $X(t)$. If a transition from state $X(t)$ to state $X(t+1)$ depends only on $X(t)$ and not on any of the previous states, i.e., if

$$P\{X(t+1) | X(t), X(t-1), \ldots, X(0)\} = P\{X(t+1) | X(t)\},$$

then the process is called a Markov process as we have already noted. If both the state variable X and the time variable t are discrete valued; the process is called a Markov Chain.

We denote $\qquad P\{X(t+1) = E_j | X(t) = E_i\}$ by $P_{ij}(t)$.

If the process is time-independent, the transition probabilities are said to be stationary and $P_{ij}(t) = P_{ij}$. We may write a one-step transition probability matrix as follows:

$$P = \begin{bmatrix} P_{11} & P_{12} & \cdots & P_{1n} \\ P_{21} & P_{22} & \cdots & P_{2n} \\ \vdots & & & \\ P_{n1} & P_{n2} & \cdots & P_{nn} \end{bmatrix}$$

where p_{ij} is the probability of going from state i to state j in one step.

The two-step probability matrix is given by P^2 since the probability of going from E_i to E_j in two steps is $\sum_{k=1}^{n} p_{ik} p_{kj}$ and, from inductive reasoning, the n-step probability matrix is P^n.

Markov processes have been used extensively in modeling voting behavior, queue size, and other phenomena. However, there are some processes which are slightly, but significantly different from those described above. In a Markov process, the length of time which a process spends in a given state is a random variable, independent of the particular state and any other state. Many common processes, however, do not have this underlying property. For example, everyone is familiar with the type of a queue which forms in front of the box office of a popular movie. Initially, when the queue is short, its size increases rapidly since people are willing to wait. However, after the queue has extended down the block and around the corner, people who arrive and wish to see the movie decide that they would rather not wait and they go elsewhere. The line then moves from the state of having a few people waiting before a box office with constant service rate, for example, to a state where the rate at which people join the queue depends on how many people are already there.

The treatment of an illness is another example of such a queue. The time which the patient spends in a particular stage of an illness is often a function of that stage and also of the stage which he is going to next.

To summarize, the salient points of these processes, called semi-Markovian processes, are as follows. The transitions between states and their probabilities are those of a Markov process, but the time spent in each stage is a random variable which is a function of both the state the process is in and the state to which the process is changing. For both Markov and semi-Markov processes, we are interested in such quantities as the expected time of first passage to state j, starting in state i, the number of visits to state j in n state changes, the number of visits to state j between visits to state i, the mean time between recurrences of state i, the mean time to absorption, given there are one or more states which absorb the process (machine failure, etc.). Absorption occurs at a state when the process is trapped there and cannot get out of that state. We illustrate the differences between the two types of process by looking at the so-called first passage time, the expected length of time for the process to get to state j for the first time, given that it starts out in state i.

Let m_{ik} be the Markov mean first passage time from i to k; l_{ik} be the semi-Markov mean first passage time from i to k; $F_{ik}(t)$ be the distribution function, for the semi-Markov process, for the "wait" of the process in state i, given that the next transition will be to state k; and h_{ik} be the mean of the distribution function $F_{ik}(t)$.

Now there are two possible ways to get from i to j: (i) directly from i to j in one transition, or (ii) from i to some state $k \neq j$ in one transition, and then from k to j in an unknown number of transitions which will take either m_{kj} or l_{kj} time units, depending on whether it is a Markov or semi-Markov process. This leads to the following sets of equations, one set for each process, which can be solved for the m_{ij} or l_{ij} respectively.

Markovian:

$$\sum_{k \neq j} P_{ik}(m_{kj}+1) + p_{ij}(1) = m_{ij}, \quad \text{for all } i.$$

Semi-Markovian:

$$\sum_{k \neq j} p_{ik}(l_{kj} + h_{ik}) + p_{ij}(h_{ij}) = l_{ij}, \quad \text{for all } i.$$

The calculations are similar, once the h_{ik}'s have been determined (in practice, this may not be a trivial task).

We give a very simple example of a Markov process: weather forecasting. Let the three types of weather in a specific city be given as:

$$N \text{ (nice)}, \quad R \text{ (rain)}, \quad S \text{ (snow)}.$$

We have been given reason to assume that the transition probability matrix from one day to the next is given by:

	N	R	S
N	$\frac{1}{2}$	$\frac{1}{4}$	$\frac{1}{4}$
$P_1 =$ R	$\frac{1}{2}$	0	$\frac{1}{2}$
S	$\frac{1}{4}$	$\frac{1}{4}$	$\frac{1}{2}$

We want to know the probabilities over a two-day period, and, and also in the long run.

Over two days, the probabilities are given by $P_2 = P_1^2$.

$$P_2 = P_1^2 = \begin{bmatrix} \frac{1}{2} & \frac{1}{4} & \frac{1}{4} \\ \frac{1}{2} & 0 & \frac{1}{2} \\ \frac{1}{4} & \frac{1}{4} & \frac{1}{2} \end{bmatrix} \begin{bmatrix} \frac{1}{2} & \frac{1}{4} & \frac{1}{4} \\ \frac{1}{2} & 0 & \frac{1}{2} \\ \frac{1}{4} & \frac{1}{4} & \frac{1}{2} \end{bmatrix} = \begin{bmatrix} \frac{7}{16} & \frac{3}{16} & \frac{3}{8} \\ \frac{3}{8} & \frac{1}{4} & \frac{3}{8} \\ \frac{3}{8} & \frac{3}{16} & \frac{7}{16} \end{bmatrix}$$

Thus, given that it is raining today, there is zero probability of rain tomorrow, but there is a 25% chance of rain the day after tomorrow.

The transition probability matrix for n days is found by $P_n = P_1^n$. Eventually P_n converges to a limit to give the long-range behavior. Denote the stationary probabilities of each type of weather by π_N, π_R, π_S. We may either seek the limit of P_n as $n \to \infty$ or we may solve the equation $\pi P_1 = \pi$, where $\pi = \begin{matrix} \pi_N \\ \pi_R \\ \pi_S \end{matrix}$ since this expresses the fact that the probabilities do not change from one day to the next.

This equation gives

$$\pi_N P_{NN} + \pi_R P_{RN} + \pi_S P_{SN} = \pi_N,$$
$$\pi_N P_{NR} + \pi_R P_{RR} + \pi_S P_{SR} = \pi_R,$$
$$\pi_N P_{NS} + \pi_R P_{RS} + \pi_S P_{SS} = \pi_S,$$

and we also have

$$\pi_N + \pi_R + \pi_S = 1.$$

In our example, this becomes

$$-\tfrac{1}{2}\pi_N + \tfrac{1}{2}\pi_R + \tfrac{1}{4}\pi_S = 0,$$
$$\tfrac{1}{4}\pi_N - \pi_R + \tfrac{1}{4}\pi_S = 0,$$
$$\tfrac{1}{4}\pi_N + \tfrac{1}{2}\pi_R - \tfrac{1}{2}\pi_S = 0,$$
$$\pi_N + \pi_R + \pi_S = 1,$$

to give

$$\pi_N = \tfrac{2}{5} \quad \pi_R = \tfrac{1}{5} \quad \pi_S = \tfrac{2}{5}.$$

This means that someone coming on a randomly chosen day has a 40 % chance for nice weather, a 20 % chance for a rainy day, and a 40 % chance for a snowy day.

We now proceed to another very general class of models, known as birth and death processes.

Example 4. Birth and Death Processes

Consider a continuous time-discrete state Markov chain. Suppose further that the following assumptions are satisfied: (i) the probability of going from state E_n to state E_{n+1} in the interval $(t, t + h)$ is $(\lambda_n h)(1 - \mu_n h) + o(h)$ where $o(h)$ is a function such that $\lim\limits_{h \to 0} \dfrac{o(h)}{h}$ $= 0$; (ii) the probability of a transition from state E_n to state E_{n-1} in the interval $(t, t + h)$ is $\mu_n h(1 - \lambda_n h) + o(h)$; (iii) the probability of a transition from state E_n to any state other than E_n, E_{n-1} or E_{n+1} in $(t, t + h)$ is $o(h)$; (iv) the probability of remaining in state E_n in $(t, t + h)$ is $(1 - \lambda_n h)(1 - \mu_n h) + o(h)$; (v) the transition probabilities are time-independent. Then the above process is called a birth–death process.

We may, for example, be interested in knowing how many people will be living in a city at some future date. Thus, we may be concerned with the probability $P_n(t)$ of having n persons (elements) in this city (system) at some time t. Our state E_n would then denote the state where there are n persons or elements.

Now, we may arrive at a system with n elements at time $t + h$ in several ways: there may be $n + 1$ elements at time t, and a death occurs in the interval $(t, t + h)$; there may be $n - 1$ elements at time t and a birth occurs in the interval $(t, t + h)$; there may be n elements at time t and neither a birth nor a death occurs in the interval $(t, t + h)$ or there may be $n \pm c, c \geqslant 2$, elements at time t, and some combination of two or more events (births and/or deaths) occurs in the interval $(t, t + h)$. We may summarise as in Table 5.2.

TABLE 5.2

No. of elements at time t	Event(s) in $(t, t + h)$	Probability of n elements at time $t + h$
n	No birth No death	$P_n(t)(1 - \lambda_n h)(1 - \mu_n h) + o(h)$ $= P_n(t)(1 - \lambda_n h - \mu_n h) + o(h)$
$n - 1$	Birth No death	$P_{n-1}(t)(\lambda_{n-1} h)(1 - \mu_{n-1} h) + o(h)$ $= P_{n-1}(t)\lambda_{n-1} h + o(h)$
$n + 1$	Death No birth	$P_{n+1}(t)(1 - \lambda_{n+1} h)(\mu_{n+1} h) + o(h)$ $= P_{n+1}(t)\mu_{n+1} h + o(h)$
$n \pm c$	More than 1 event	$o(h)$

Thus

$$P_n(t + h) = \lambda_{n-1} h P_{n-1}(t) + (1 - \lambda_n h - \mu_n h) P_n(t) + \mu_{n+1} h P_{n+1}(t) + o(h) \quad (n \geqslant 1),$$
$$P_0(t + h) = (1 - \lambda_0 h) P_0(t) + \mu_1 P_1(t) + o(h),$$

or

$$P_n(t + h) - P_n(t) = \lambda_{n-1} h P_{n-1}(t) - (\lambda_n + \mu_n) h P_n(t) + \mu_{n+1} h P_{n+1}(t) + o(h).$$

Therefore,

$$\lim_{h \to 0} \frac{P_n(t+h) - P_n(t)}{h} = \lambda_{n-1} P_{n-1}(t) - (\lambda_n + \mu_n) P_n(t) + \mu_{n+1} P_{n+1}(t),$$

i.e.,

$$\frac{dP_n(t)}{dt} = \lambda_{n-1} P_{n-1}(t) - (\lambda_n + \mu_n) P_n(t) + \mu_{n+1} P_{n+1}(t).$$

If we assume that all elements in the population have an equal probability of giving birth to a new element, and similarly for deaths, then $\lambda_n = n\lambda$ and $\mu_n = n\mu$.
Thus

$$\frac{dP_n(t)}{dt} = (n-1)\lambda P_{n-1} - (n\lambda + n\mu) P_n(t) + (n+1)\mu P_{n+1}(t).$$

A process may be absorbed or trapped in a state or it may be reflected to a neighboring state. Thus E_N is an absorbing barrier if $\lambda_n = \mu_n = 0$ for $n \geq N$. The state E_0 which now is a reflecting state becomes an absorbing state if $\lambda_0 = 0$. Applications of these ideas have been made to the spread of epidemics.

The following is an illustration of embedded Markov chains in a birth–death context.

Example 5. The Pollaczek–Khintchine Queue [Saaty]

Suppose that arrivals occur at random by a Poisson process with the rate λ per unit time, to a waiting line, in statistical equilibrium, before a single-service facility. Also suppose that they are served by an arbitrary service-time distribution at the rate of μ per unit time— first come, first served. We assume that $\lambda/\mu < 1$. Suppose that a departing customer leaves q in line, including the one in service, whose service time is t. Let r customers arrive during this time t. If the next departing customer leaves q' customers behind, we can relate q and q' as follows:

$$q' = \max(q-1, 0) + r = q - 1 + \delta + r,$$

where

$$\delta(q) = \begin{cases} 0 & \text{if } q > 0, \\ 1 & \text{if } q = 0. \end{cases}$$

Note that by introducing δ we avoid using the max expression.

It is assumed that equilibrium values for the first and second moments $E[q]$ and $E[q^2]$ of the queue exist. Note that q is treated as a random variable. Now, from the definition we have, $\delta^2 = \delta$ and $q(1 - \delta) = q$. Also, $E[q] = E[q']$ and $E[q^2] = E[q'^2]$, since both q and q' are assumed to have the same equilibrium distribution. Note that, because the system is in equilibrium there is no difference between the types of queue left behind any customer; i.e., they all have the same probability distribution independent of time.

Taking the expected value of the equation in the different variables, we have

$$E[q'] = E[q] - E[1] + E[\delta] + E[r]$$

from which we have

$$E[\delta] = 1 - E[r].$$

But during a service time of length t we have

$$E[r|t] = \sum_{r=0}^{\infty} r \frac{(\lambda t)^r}{r!} e^{-\lambda t} = \lambda t, \quad E[r] \equiv \int_0^\infty E[r|t] f(t) dt = \frac{\lambda}{\mu},$$

$$E[r^2|t] = \sum_{r=0}^{\infty} r^2 \frac{(\lambda t)^r}{r!} e^{-\lambda t} = (\lambda t)^2 + \lambda t.$$

Thus, on taking expectations with respect to the service time t, one has $E[r] = \lambda/\mu \equiv \rho$. For example, if the service distribution is exponential, we have

$$E[r] = \mu \int_0^\infty (\lambda t) e^{-\mu t} \, dt = \frac{\lambda}{\mu} \equiv \rho$$

since the average value of the service-time distribution is $1/\mu$.

Note that $E[r]$ is a number which is unaffected by taking averages. This gives $E[\delta] = 1 - \rho$. Now the probability of r arrivals is independent of q, the length of the queue, and of δ, which assumes values that depend only on q which is independent of r. Consequently, if we take the expected value over the variables r, q, and δ, we may take the expected value of r and of q separately wherever we encounter their product. This is also true for r and δ.

Also, averaging r^2 over time yields

$$E[r^2] = \lambda^2 \operatorname{var}(t) + \rho^2 + \rho,$$

$$\operatorname{var}(t) \equiv \int_0^\infty \left(t - \frac{1}{\mu} \right)^2 f(t) \, dt.$$

Again, by squaring the equation relating q and q' and using the facts that $\delta^2 = \delta$, $\delta q \equiv 0$,

$$q'^2 = q^2 - 2q(1-r) + (r-1)^2 + \delta(2r-1).$$

Hence, because of equilibrium,

$$0 = E[q'^2] - E[q^2] = 2E[q]E[r-1] + E[(r-1)^2] + E[\delta]E[2r-1]$$

or, simplifying and using the foregoing relations, we have the Pollaczek–Khintchine formula:

$$\begin{aligned}
E[q] &= \frac{E[(r-1)^2] + E[\delta]E[2r-1]}{2E[1-r]} \\
&= \frac{E[r^2] - 2E[r] + 1 + E[\delta](2E[r]-1)}{2(1 - E[r])} \\
&= \frac{\lambda^2 \operatorname{var}(t) + \rho^2 + \rho - 2\rho + 1 + (1-\rho)(2\rho - 1)}{2(1-\rho)} \\
&= \rho + \frac{\rho^2 + \lambda^2 \operatorname{var}(t)}{2(1-\rho)}.
\end{aligned}$$

Thus, once we know the variance of the service time t from its given distribution, the average number in the system is determined. It is important to observe that the foregoing average is taken over instants just following departures and is not the time average value of the number in the system. In fact, if $E_t(q)$ is the time average, all we can say without further argument is that $E[q] \leqslant E_t(q) < E[q] + 1$.

Note the fact, which holds in general (i.e., even if one has several channels in parallel), that the average number in the system equals the sum of the average number of busy channels (in this can it is ρ, the traffic intensity) and the average number in line.

Now to obtain the average waiting time we argue as follows: If we write $E[w]$ for the average waiting time in the queue (not including service), $\lambda[E(w) + 1/\mu]$ is the expected number of arrivals during the total waiting plus service time of one customer; i.e., of his stay in the system. But this must be just the number in the system immediately after his departure; namely $E[q]$. Thus

$$W_q \equiv E[w] = \frac{\rho^2 + \lambda^2 \, \text{var}(t)}{2\lambda(1 - \rho)} = \frac{L_q}{\lambda}.$$

The smaller $\text{var}(t)$ the less is the waiting time, e.g., if the service distribution is constant we have $\text{var}(t) = 0$. Note that if we consider the birth–death equations and put $\lambda_n = \lambda$ and $\mu_n = \mu, \mu_0 = 0$, and equate the derivatives to zero, we obtain the steady-state equations of a single-channel first-come first-served queue with Poisson input and exponential service time:

$$0 = -(\lambda + \mu)p_n + \lambda p_{n-1} + \mu p_{n+1} \quad (n \geqslant 1),$$
$$0 = -\lambda p_0 + \mu p_1,$$

from which we have

$$p_n = \rho^n(1 - \rho), \; \rho \equiv \frac{\lambda}{\mu} < 1, \text{ and}$$

$$L_q = \sum_{n=0}^{\infty} np_n = \frac{\rho}{1 - \rho}$$

for the number in queue. Check the Pollaczek–Khintchine formula for this result by putting $f(t) = e^{-\mu t}$.

Another general process of wide application follows.

Example 6. Gaussian (Normal) Process

The importance of the Gaussian distribution in probability theory arises from the fact that many random variables may be considered as normally distributed and the normal distribution is tractable and convenient to handle.

A stochastic process $\{x(t), t \in T\}$ is said to be a Gaussian process if, for any integer n and any subset $\{t_1, t_2, \ldots, t_n\}$ of T, the n random variables $X(t_1), X(t_2), \ldots, X(t_n)$ are jointly normally distributed. An example of a Gaussian process is the Wiener process which also provides a mathematical model for Brownian motion.

Consider the motion of a particle on a line as a result of numerous random impacts with other particles. Take the origin as its position at time $t = 0$. If we assume that the time between successive impacts are independently and exponentially distributed with mean $1/\mu$, then the number $N(t)$ of impacts in time t is a Poisson process, with intensity μ. Assume that the effect on the particle by an impact is to change its position by $\pm a$, each having a probability of $\frac{1}{2}$. The position of the particle at time t may be represented by $X(t) = \sum_{n=1}^{N(t)} Y_n$, where Y_n is the particle's change of position resulting from the nth impact. It can be shown that $\{X(t), T > 0\}$ is a stochastic process with stationary independent increments. Its characteristic function is

$$\phi_{X(t)}(x) = \exp(\mu t \ E[\exp(ix Y) - 1]\}$$
$$= \exp[-(1/2)x^2 a^2 \mu t + \theta |x|^3 a^3 \mu t].$$

Defining $\sigma^2 = a^2 \mu$, the logarithm of the characteristic function is given by

$$\log \theta_{X(t)}(x) = -\frac{1}{2}x^2\sigma^2 t + a\theta|x|^3\sigma^2 t.$$

Now let $\mu \to \infty$ and $a \to 0$ in such a way that $\mu a^2 = \sigma^2$ (the total mean square displacement of the particle per unit time) is constant. Then, $\phi_{X(t)}(x) \to \exp(-\frac{1}{2}x^2\sigma^2 t)$.

Thus, $X(t)$ is approximately normally distributed with stationary independent increments. This is the Wiener process, which is a special case of the Gaussian process with $E[X(t)] = 0$.

The methodology has been applied to noise currents in a vacuum tube.

If each element of a system produces a number of new elements at each stage of a process according to a probability law, we have a growth phenomenon known as a branching process.

Example 7. Bacterial Growth: A Branching Process

For example, consider bacteria which reproduce only at discrete intervals. Let p_k be the probability that a single organism reproduces k new organisms in one time period. We shall assume that all organisms reproduce independently of the others. In this particular problem, we make the simplifying assumption on X_t, the number in the system at time t, that $X_0 = 1$. Then X_1 has the probability distribution $\{p_k\}$ and $P(s) = \sum_{k=0}^{\infty} p_k s^k$ is the probability generating function of X_1. The generating function of X_2, $P_2(s)$, is given by

$$P_2(s) = \sum_{k=0}^{\infty} p_k[P(s)]^k$$
$$= P[P(s)].$$

If $X_1 = n$, $P_2(s|X_1 = n) = [P(s)]^n$.
Similarly,

$$P_n(s) = P[P_{n-1}(s)].$$

From this we can derive the extinction probabilities $P(X_t = 0)$ for each state. If $p_0 = 0$, the extinction probabilities are obviously zero, so we shall assume that $0 < p_0 < 1$. If x_t is the extinction probability at the tth stage, then $x_t = P(X_t = 0)$.

Consider the probability generating function

$$P_t(s) = \sum_{n=0}^{\infty} P(X_n = n)s^n$$

$$= P(X_t = 0)s^0 + \sum_{n=1}^{\infty} P(X_t = n)s^n.$$

We have

$$P_t(0) = P(X_t = 0) = x_t.$$

Further,

$$x_1 = p_0, \quad x_2 = \sum_{n=0}^{\infty} p_n p_0^n$$

since all organisms act independently. Recursively, we have

$$x_t = P(x_{t-1}).$$

Since $P(s)$ is a monotone increasing function for $0 < s < 1$, we have $x_1 < x_2 < \ldots < x_n < \ldots$.

Further, $P(s)$ increases to some number τ such that $\tau = P(\tau)$. If u is any positive root of $u = P(u)$, then $x_n < P(u) = u$ for all n and hence, $\tau < u$. Therefore, $\lim_{t \to \infty} x_t =$ the smallest possible root of $u = P(u)$. Now, $u = P(u)$ has a positive root less than unity, if and only if $P'(1) > 1$ since $P(s)$ is a convex monotone function.

Let $\mu = \sum_{n=0}^{\infty} n P_n = P'(1)$ be the expected number of descendants of a particular organism. Then if $\mu < 1$, the probability of eventual extinction tends to one; that is, $\lim_{t \to 0} x_t = 1$. If $\mu > 1$, there exists some $\xi < 1$ such that $\lim_{t \to \infty} x_t = \xi$, which is the probability of extinction in finite time. Further, we call $1 - \xi$ the probability of an infinitely prolonged process, or of the establishment of the process.

Now if $X_0 = r$, then $\lim_{t \to \infty} x_t = \xi^r$ if $u > 1$ and the probability of establishment is $1 - \xi^r$, which is near 1 if r is large. Further, by induction, we obtain the result $E(x^r) = \mu^r$. Thus, either the process explodes or expires, and a stable position is highly improbable.

Example 8. Availability

Let $F(t)$ be the failure density function at time t of a component in an electronic system and $P(t)$ the cumulative probability that maintenance will not be completed by time t. The unavailability $U(t)$ of the component at time t given that it was put into operation at time $t = 0$ is given by

$$U(t) = \int_0^t F(t) P(t) dt.$$

Let

$$U = \lim_{t \to \infty} U(t) = \int_0^{\infty} F(t) P(t) dt.$$

We define the availability $A(t)$ as

$$A(t) = 1 - U(t), \quad \text{and let} \quad A = 1 - U$$

If

$$F(t) = \lambda e^{-\lambda t}, \; P(t) = \int_t^{\infty} g(t) dt = e^{-\mu t} \; M = \frac{1}{\lambda}, \; D = \frac{1}{\mu},$$

then
$$A(t) = \frac{M}{M+D} + \frac{D}{M+D} \exp\left[-\left(\frac{1}{M}+\frac{1}{D}\right)t\right]$$

and hence
$$A = \frac{M}{M+D}.$$

The same steady-state result can be obtained by using arbitrary distributions for $F(t)$ and $P(t)$. [Note that if we defined availability A as a random variable $A = X/(X+Y)$, which is a ratio of the time between failures X and the time between failures X plus the time needed for replacement Y, then the expected value of A cannot be written as the ratio of the expected values. If Y/X is always very small we may expand the right side in powers of Y/X as follows:

$$A = \frac{1}{1+Y/X} = 1 - Y/X + (Y/X)^2 - \cdots,$$

and then take the expected value on both sides. The last term of the alternating series provides an estimate of the error.]

In our next chapters we consider a number of examples classified by area of application rather than by type of model.

Chapter 5—Problems

1. A shot is fired from each of three guns A, B, C at a single target, the probability of a hit being 0.1, 0.2, 0.3 respectively. Find the probability distribution for the number of shots.

2. In the circuit of Fig. 5.1 (p. 81), given that the bulb is lit, what is the probability that switches A and B are both closed? $\left[\frac{4}{13}\right]$.

3. If two integers A and B, $B > A$ are selected at random, what is the probability that they have no common divisor?

$$\left[P = \left(1 - \frac{1}{2^2}\right)\left(1 - \frac{1}{3^2}\right)\left(1 - \frac{1}{5^2}\right) = \frac{6}{\pi^2}.\right]$$

4. A player X plays one game with one of two players A and B, a second game with the other and finally, a third game with the first one. The probability that X wins a game against A is $1/3$; that against B is $2/3$. X wins if he wins two consecutive games. Should he play the sequence ABA or BAB to improve his chances of winning? [ABA.]

5. Consider an organism whose only purpose is searching for food on a branching system of paths. It has no maximization objective but only needs to maintain a certain average rate of food intake (a satisficing objective). The food occurs at random on any of the nodes of the branching system. The probability that food is located at any node is $p > q$. From any node there are d branches to d nodes. The organism can see all nodes that are n moves away from its position and hence can follow a definite path if food is seen ahead. (A move is a transition along a branch from one node to an adjacent one.) If food has not been sighted, the organism is indifferent as to which node to approach next. The maximum number of moves the organism can make between meals without starvation is N. Determine P, the probability that the organism will survive from meal to meal.

$$P = 1 - (1-p)^{(N-n)d^n}.$$

6. A needle of length l is thrown at random on a board on which parallel lines are drawn d units apart. If $d > l$, what is the probability that the needle touches one of the parallel lines? $\left[\dfrac{2l}{\pi d}.\right]$

7. In a study of past records, it has been found that 25% of all shirts manufactured had an imperfection in them. Two persons are hired to inspect the shirts before they are shipped from the factory. The probability that either inspector will misclassify a shirt is 10% and their decisions are independent. (a) If it is decided to class as imperfect any shirt which either or both inspectors reject, what is the probability that if a shirt is classified as good, it is actually good; actually imperfect? (b) If it is decided to class as imperfect only the shirts that both inspectors reject, what is the probability that a shirt classified as good is actually good; actually imperfect? Also, if it is classified as imperfect, what is the probability it is actually good; actually imperfect? [*Hint*: use Bayes' theorem.]

Bibliography

Bailey, N. T. J., *The Elements of Stochastic Processes With Applications to Natural Sciences*, Wiley, New York, 1964.

Bartlett, M. S., *Stochastic Processes*, Cambridge University Press, 1962.

Benes, V. E., *General Stochastic Processes in the Theory of Queues*, Addison-Wesley, Reading, Massachusetts, 1963.

Bharucha-Reid, A. T., *Elements of the Theory of Markov Processes and Their Applications*, McGraw-Hill, New York, 1960 (Chapter 9, The Theory of Queues).

Bodino, G. A., and F. Brambilla, *Teoria delle Code*, Cisalpino, Milano, 1959.

Chung, K. L., *Markov Chains with Stationary Transition Probabilities*, Springer-Verlag, Berlin, 1960.

Cox, D. R. and W. L. Smith, *Queues*, Wiley, New York, 1961.

Doob, J. L., *Stochastic Processes*, Wiley, New York, 1953.

Dynkin, E. B., *Markov Processes* (2 volumes), Academic Press and Springer-Verlag, 1965.

Dynkin, E. B., *Theory of Markov Processes*, Prentice-Hall, Englewood Cliffs, New Jersey, 1961.

Feller, W., *An Introduction to Probability Theory With Applications*, Vol. II, Wiley, New York, 1966.

Feller, W., *An Introduction to Probability Theory and Its Applications* (2nd edn.), Wiley, New York, 1957.

Girault, M., *Stochastic Processes*, Springer-Verlag, Berlin, 1966.

Harris, T. E., *The Theory of Branching Processes*, Prentice-Hall, Englewood Cliffs, New Jersey (Springer-Verlag), 1963.

Howard, R. A., *Dynamic Programming and Markov Processes*, Wiley, New York, 1960.

Ito, K. and H. J. McKean, *Diffusion Processes and their Sample Paths*, Springer-Verlag, Berlin, 1965.

Karlin, S., *A First Course in Stochastic Processes*, Academic Press, New York, 1968.

Kemeny, J. G., and J. L. Snell, *Finite Markov Chains*, van Nostrand, New Jersey, 1960.

Kemperman, H. B., *The Passage Problem for a Stationary Markov Chain*, University of Chicago Press, Chicago, 1961.

Khintchine, A. Y., *Mathematical Methods in the Theory of Queueing*, Griffin, London, 1960.

LeGall, P., *Les Systemes Avec ou sans Attente*, Editions Dunod, Paris, 1962.

Loeve, M., *Probability Theory*, 3rd edn., van Nostrand, New Jersey, 1963.

Meyer, P. A., *Probability and Potentials*, Blaisdell, New York, 1966.

Morse, P. M., *Queues, Inventories and Maintenance*, ORSA, Wiley, New York, 1958.

Neven, T., *Mathematical Foundations of the Calculus of Probability*, Holden-Day, San Francisco, California, 1965.

Parzen, E., *Stochastic Processes*, Holden-Day, San Francisco, California, 1962.

Pollaczek, F., *Theorie Analytique des Problems Stochastiques Relatifs a un Groupe de Lignes Telephoniques avec Dispositif d'attente*, Mem. Sci. Math., Gauthier-Villars, Paris, 1961.

Pollaczek, F., *Problemes Stochastiques Poses par le phenomene de Formation d'une queue d'attente a un Guichet et par des Phenomenes Apparentes*, Mem. Sci. Math., Vauthier-Villars, Paris, 1957.

Prabhu, N. U., *Queues and Inventories*, Wiley, New York, 1965.

Prabhu, N. U., *Stochastic Processes*, Macmillan, New York, 1965.

Riordan, J., *Stochastic Service Systems*, Wiley, New York, 1962.

Rosenblatt, M., *Random Processes*, Oxford University Press, Oxford, 1962.

Runnenburg, J., Th., *On the Use of Markov Processes in One Server Waiting Time Problems and Renewal Theory*, University of Amsterdam, 1960.

Saaty, T. L., *Elements of Queueing Theory With Applications*, McGraw-Hill, New York, 1961.

Spitzer, F., *Principles of Random Walk*, van Nostrand, New Jersey, 1964.

Syski, R., *Introduction to Congestion Theory in Telephone Systems*, Oliver & Boyd, Edinburgh, 1960.

Syski, R., Stochastic differential equations, Chapter 8, in T. Saaty, *Modern Non-linear Equations*, McGraw-Hill, New York, 1967.

Takacs, L., *Introduction to the Theory of Queues*, Oxford University Press, Oxford, 1962.

Takacs, L., *Stochastic Processes* (*Problems and Solutions*), Methuen, London, 1960.

PART III

APPLICATIONS

WE have attempted to give a useful but greatly oversimplified division of the various areas of application of mathematics into the physical, biological, social, and behavioral sciences. Of course, many disciplines are hybrids, developed from a combination of these sciences—for example, biochemistry which is a hybrid of the biological and physical disciplines. This suggests using the following table or rectangular array to classify applications. We have mentioned only a small number of the existing hybrid areas; some of them may fall in a square that is different from the one the reader thinks appropriate. In the four chapters of this part of the book, we give applications in both the basic and hybrid areas shown.

The following chart is NOT symmetric since we adopt the convention that rows dominate columns; that is, genetics is more biological than physical, while biomedical engineering is more physical than biological.

The next three chapters are concerned with the physical and biological, social and behavioral sciences, and with decision problems.

	Physical	Biological	Social	Behavioral
Physical	Physics Chemistry Astronomy Meteorology Geology	Biomedical engineering Physiology	Technology Mathematical sociology (statistics)	Control theory Queuing theory General systems theory
Biological	Medicine Genetics Biochemistry Molecular biology Biophysics	Biology Disease	Ecology Epidemiology	Psychology
Social	Physical Anthropology History	Pollution	Sociology Anthropology	Social psychology Organization theory communication
Behavioral	Economics	Imprinting Behavioral physiology (aggression)	Conflict resolution Political science	Management and behavioral science

Chapter 6

Physical and Biological Applications

6.1. Introduction

IN this chapter we give examples of models in areas with emphasis on the physical and biological aspects of modeling.

Some of the earliest applications of mathematical models have occurred in the physical sciences. It is by reference to such applications that Eugene Wigner, the well-known physicist, has spoken of the "unreasonable effectiveness of mathematics in the natural sciences." Much of that mathematics has a hard core of calculus. As the applications are extended to the biological and social and behavioral sciences, there is an attempt to use the same tools, but they do not yet seem to be as effective in the analysis of these subjects as they have been in the natural sciences.

6.2. Natural Sciences

We shall consider examples from a number of areas in the natural sciences.

Example 1. Models in Astronomy—The historical importance of information

One area of science in which the use of mathematical models has been essential is the field of astronomy and its cousins astrophysics and cosmology. The problem with astronomy is that observations are necessarily indirect. We must analyze the visual light, radio waves, X-rays, and γ-rays which were emitted from celestial objects millions of years ago because we are subject to the enormous time lags produced by the snail-like pace with which light crawls across the cosmos.

Let us take a case in point. The presence of various elements in the atmosphere of a star induces the absorption of corresponding wavelengths from the light which the star radiates, and this can be observed by the existence of dark bands at these specified wavelengths in the spectrum of the star. Similarly, such lines are observed in the spectra of entire galaxies.

The *red shift* (an astronomical, not political, phenomenon) is simply that all these lines occur slightly more toward the red end of the spectrum than one would expect. (The location of these lines depends on the amount of energy involved in the transition of an electron from one quantum state to another and so is precisely determined.) Furthermore, it appears that the farther away the object observed, the more pronounced its red shift.

There are several explanations which help account for this phenomenon. One holds that the entire universe is expanding and that the red shift is nothing more than an optical

Doppler effect. The galaxies are visualized as lying on the three-dimensional "surface" of the universe like points on the surface of an expanding balloon. (The Doppler effect is the rising in pitch of an approaching train whistle or police siren and the analogous falling in pitch when the locomotive or police car is receding.) Other suggestions include relativistic aspects and also the existence and density of intergalactic dust and gas.

Obviously, the astronomer's task is very difficult. In fact, given essentially more variables than equations, how does he get anywhere? The technique followed by a successful astronomer is really just that of a good detective (or, for that matter, a good bridge player). He guesses a lot and then tries to confirm or refute his hypotheses.

Analysis of the exact causes of the red shift is very important to cosmology—the theory of the nature of the universe. Astronomers hope to be able to announce a winner in the race between the "big-bang," "steady-state," and "pulsating-universe" theories by studying the red shift of quasars, which are intensely bright and incredibly distant celestial objects scattered fairly evenly throughout the known universe. It might also be possible to determine what the shape and size of the universe is by such observations.

In 1914 the American astronomer Vesto Melvin Slipher presented before the American Astronomical Society at Evanston, Illinois, a paper with slides clearly showing the rapid movement of galaxies as indicated by the red shift. In 1916 the Dutch astronomer Willem de Sitter found a solution to Einstein's general relativity equations that predicted an expanding universe in which the galaxies moved rapidly away from each other. A similar solution was later discovered by the Russian mathematician Alexander Friedmann by finding an error in Einstein's algebra! (Einstein had divided by zero.) In 1929 Edwin Hubble formulated as a result of experimental work at the Mount Wilson Observatory, the law named after him: "The farther away a galaxy is, the faster it moves."

In 1965 Arno Pinzias and Robert Wilson of the Bell Laboratories discovered that the earth is bathed in a faint glow of radiation coming from every direction in the heavens.

These discoveries have tended to confirm the big-bang theory that the universe started from one dense mass and spread outward by a cosmic explosion (or creative act), 20 billion years ago, first forming a white-hot fireball. Its radiation would never have disappeared, and that is the measurable radiation discovered at the Bell Labs. Most of the laws of nature may have meaning only in the new expanded universe.

In physics, gravitation, motion, light, space, and time have so far served as basic entities to explain observed mechanical and relativistic occurrences within natural law. However, the interaction of elementary particles has needed a new unifying master theory to combine the four forces observed to operate in the background. They are gravitation and electromagnetism with an unlimited range of influence and two other forces which cannot be perceived directly because of a range of influence limited to atomic nucleii. One is the strong force which binds together the protons and neutrons (and their constituent quarks) in the nucleus. The other is the weak force responsible for the decay of certain particles.

Among the most interesting events in recent geological history are the glaciations which have periodically inundated large portions of the earth in a torrent of ice. Many explanations have been advanced. Some scientists ascribe glacial epochs to variations in the composition of the earth's atmosphere, i.e., more or less carbon dioxide or dust. (This explanation, whether or not true in the past, may well apply to the future of an increasingly polluted planet.)

Other speculations focus on more astronomical causes. Periodic sunspot activity is known to affect average terrestrial temperature, decreasing it by about 2°F. Other variations in solar activity could account for more significant temperature fluctuation.

Finally, we can investigate the influences of the earth's orbit and rotation, the movement of the earth's axis which rotates to form a cone over a period of about 26,000 years, and variations in the eccentricity of the orbit. Using the Stefan–Boltzmann formula

$$\frac{T_1}{T_2} = \sqrt[4]{L_1/L_2},$$

where L_1 and L_2 denote the amount of heat received and T_1 and T_2 represent corresponding surface temperature in degrees centigrade above absolute zero, we can compute the temperature of the earth subject to astronomical fluctuations. In fact, calculations of periods of minimum terrestrial temperature, using only these astronomical factors, have corresponded very well with periods of glaciation.

Note, however, that some geological input is still necessary, for there is evidence that the present sequence of glaciations does not extend back indefinitely. Geologists explain this by pointing out that only during mountainous stages of the earth's history (and we are still in one now) are the conditions right for formation of glaciers during increasingly cold periods.

Our next example is from the field of chemistry.

Example 2. Markov Chains and Chemical Processes

The following example shows how matrices and Markov chains can be used to study first-order chemical processes. Consider a hypothetical reaction $A \longrightarrow B$ and assume that at any time during the reaction the probability that a molecule A changes to a molecule B is 0.1. The probability that a molecule A will remain unchanged during the same period of time is 0.9. Assuming that the reverse relation of $B \longrightarrow A$ is not allowed, the probabilities describing the transitions of the system may be represented by the matrix

$$\begin{array}{cc} & \begin{array}{cc} A & B \end{array} \\ \begin{array}{c} A \\ B \end{array} & \begin{pmatrix} 0.9 & 0.1 \\ 0 & 1 \end{pmatrix} \end{array}$$

Suppose that the initial state is $A = 1$ and $B = 0$, then the state after a single transition will be $(1, 0) \begin{pmatrix} 0.9 & 0.1 \\ 0 & 1 \end{pmatrix} = (0.9, 0.1)$ and the state after a second transition will be

$$(0.9, 0.1) \begin{pmatrix} 0.9 & 0.1 \\ 0 & 1 \end{pmatrix} = (0.81, 0.19)$$

or

$$(1, 0) \begin{pmatrix} 0.9 & 0.1 \\ 0 & 1 \end{pmatrix} \begin{pmatrix} 0.9 & 0.1 \\ 0 & 1 \end{pmatrix} = (1, 0) \begin{pmatrix} 0.9 & 0.1 \\ 0 & 1 \end{pmatrix}^2$$

and so on until we reach the tth transition. This can be generalized by replacing the probability 0.1 by k and allowing t, the number of transitions, to represent the number of units of time. The probable state of the system after t transitions is given by $(1, 0) \begin{pmatrix} 1-k & k \\ 0 & 1 \end{pmatrix}^t.$

For a system of N molecules A, the fraction of molecules that remain like A after t transitions is

$$\frac{n}{N} = (1 - k)^t.$$

If k is small and t is large, we can approximate this by

$$\frac{n}{N} = (1 - k)^t = e^{-kt}$$

from which we have $kt = -\ln \frac{n}{N}$. This is a well-known equation for a first-order process in which during a time of length t_0 the probability of transition from A to B is kt_0.

The same approach can be used in reactions involving more than two types of molecules or in reversible reactions.

6.3.   Oxygen and Blood Circulation

Not too long ago, engineers began to look at the human body as a machine, and the science of biomedical engineering was born. Because of their novelty and the imagination used in their formulation, we shall give a number of models developed in this area. Several of these examples have been developed for the field of robotics (the design of machines capable of carrying out human functions).

As an example of a machine which functions *in vivo*, we might cite the pacemaker, while the artificial kidney is much too large for artificial implantation. Both machines perform vital bodily functions for people whose bodies need assistance at these tasks. Indications are certainly that the present trend will continue. For example, one goal of current research is to design a machine which will monitor the blood–sugar level in diabetics and automatically administer an appropriate dose of insulin. (In fact, the machine has already been developed and successfully tested on animals, but it is not yet small enough to be implanted or worn by a diabetic.) The ultimate in such machines would be the "medikit" of Murray Leinster or the "autodoc" of Larry Niven, devices which will be familiar to science-fiction devotees.

We now give some simple examples.

Example 1. Diffusion in the Body

The process by which a gas or liquid spreads throughout another medium is called diffusion. We give two direct applications of the principles of diffusion of gases and liquids in the body; the first application, by Danziger and Emergreen, develops a simple model of the endocrine system; the second application, by Defares and Sneddon, analyzes the oxygen debt in the body after exercise.

Danziger and Emergreen consider a system of n components whose concentrations, $x_1 \ldots x_n$, in the body are functions of time. This leads to the system of equations

$$\frac{dx_i}{dt} + \lambda_i x_i = Q_i, \quad (i = 1, 2, \ldots, n),$$

where λ_i is the loss rate per unit concentration of component i and Q_i is the production rate of x_i. This production rate may be approximated by

$$Q_i = A_{i0} + \sum_{j=1}^{n} A_{ij}x_j \quad (i = 1, 2, \ldots, n),$$

where A_{i0} is zero or a positive constant denoting independent production of x_i, the A_{ij} are sensitivity constants which may be zero for no effect, positive for production stimulation, or negative for production inhibition, and $A_{ii} = 0$.

Defares and Sneddon explain how the measurement of oxygen debt is used as a heart function test. During strenuous exercise the body can withstand an oxygen debt x (the muscles are able to work without oxygen supply θ) but in the recovery stage oxygen is needed to replenish the energy stores used up during exercise. They assume that the oxygen debt is proportional to the work W which is done, i.e., $x = \alpha W$, and also that $dx/dt - d\theta/dt = \alpha dW/dt$, i.e., the oxygen debt also depends upon the oxygen supply. Finally, we assume that the extra oxygen uptake per second (at the lungs) is proportional to the oxygen debt existing at any instant, $d\theta/dt = kx$, to obtain

$$\frac{1}{k}\frac{dx}{dt} + x = \frac{\alpha}{k}\frac{dW}{dt} = \frac{\alpha}{k}P(t).$$

Various functional forms may be chosen for $P(t)$ in order to solve for x; this allows $d\theta/dt = f(t)$ to be calculated. $f(t)$ can be measured to provide a check on the model.

We now describe some physical models of the lung.

*Example 2. Models of the Lung

The lung functions in a variety of ways. It acts as a bellows pumping air in and out of the body, but it also includes a membrane which permits the osmosis of oxygen into the blood. Different models of the lung may reflect or emphasize these distinct but interrelated functions.

The model developed by Collins, Kilpper, and Jenkins is based on a physical analog of a piston and springs and is used to analyze the mechanical factor in respiration.

They created a physical analog of the lung thorax system as shown in Fig. 6.1. They explain that the gravitational and active muscular forces on the thorax give rise to a pressure P_x. These forces are equivalent to an external pressure acting inward on the thorax in addition to the atmospheric pressure P_0. Acting outward is the pressure P_p in the fluid lubricated pleural space. Since the algebraic sums of these forces are balanced by the stress due to the elastic property of the thorax, it makes sense to write, after some reflection,

$$P_x + P_0 - P_p = -1/C_T\{V - EV_T\},$$

where V is the lung alveolar volume, EV_T is the equilibrium volume the lungs would assume if the elastic forces of the thorax were the only forces acting, and C_T is the compliance of the thorax.

A similar equation describes the equilibrium of forces on the lungs:

$$P - P_p = 1/C_L\{V - EV_L\},$$

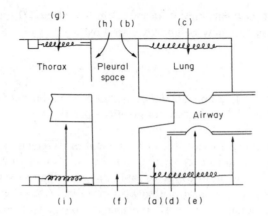

FIG. 6.1. Single-compartment model of a lung-thorax system: (a) lung volume V; (b) piston represents pleural wall of lungs; (c) springs represent lung compliance C_1; (d) lower airway resistance R_1 varies with lung volume; (e) intermediate and upper airway resistance R_2 may vary with pressure; (f) fluid-filled pleural space; (g) springs represent compliance of thorax C_T; (h) piston represents pleural wall of thorax; (i) force of shaft corresponds to muscular and gravitational force on thorax P_x.

where P is the alveolar air pressure in the lungs, EV_L is the equilibrium volume the lungs would assume if the elastic force of the lungs were the only force acting, and C_L is the compliance of the lungs.

The air flow from the lungs is accompanied by a pressure gradient along the bronchial airways. Consideration of the total pressure drop across a network of tubular airways gives rise to

$$P - P_m = R_1 \frac{dQ}{dt} + R_2 \cdot \left| \frac{dQ}{dt} \right| \frac{dQ}{dt},$$

where R_1, R_2 are resistance parameters, P_m is the pressure at the mouth, and dQ/dt is the volume flow rate of air from the lungs measured at mouth pressure.

Finally, consideration of the compressible nature of air using the ideal gas law and assuming isothermal conditions leads to the equation

$$\frac{d(PV)}{dt} = -P_m \frac{dQ}{dt}.$$

The above equations form a system of differential equations which constitute the basic description of the lung model.

We now look at the flow of blood as a physical system.

Example 3. Blood Flow

The flow of blood in the body can be likened to the flow of liquid in a system of pipes. We can develop some surprisingly accurate *a priori* bounds for the radius of the aorta.

The turbulence may be measured by the Reynolds number R, which should not exceed the value 1100.

Let

$$R = \frac{Vr\delta}{n}, \tag{6.1}$$

a measure of turbulence in the aorta, where V is the mean velocity of blood flow, r is the radius of aorta, δ is the blood density, and n is the viscosity of blood.

If C is the average blood flow in cm^3 sec^{-1}, then

$$V = \frac{C}{\pi r^2}.$$

From (6.1) we obtain for the critical (or minimal) radius

$$r^* = \frac{R^* n}{V\delta} = \frac{\pi r^{*2} n R^*}{C\delta}$$

or

$$r^* = \frac{C\delta}{1100\,\pi n}.$$

As an example, for a dog with $C\delta = 40$ g sec^{-1} and $n = 0.0269$ g cm^{-1} sec^{-1}, $r^* = 40/1100\,\pi n = 0.43$ cm, while the actual value is $r = 0.5$ cm.

*6.4.   Medical Application

We give a simple medical application concerning X-rays.

Example. Constructing Three-dimensional X-ray Pictures

X-ray photographers are two-dimensional and from one photograph one cannot infer the relative depths of the objects shown in the photograph. Suppose we take photographs from many different angles; is it possible to reconstruct reliably the X-ray density of any point in the body we are photographing? Although the solution to this involves mathematics which is not quite trivial, it will be seen that the mathematics is itself of practical importance. (This example was communicated to Saaty by Peter Bunemann.)

First of all consider a parallel X-ray beam running parallel to the x-axis penetrating a body whose X-ray density is given by the function $\rho(x, y, z)$. The energy absorbed in a small thickness dx of the body gives us

$$I(x + dx, y, z) - I(x, y, z) = -I(x, y, z)\rho(x, y, z)dx,$$

where I is the intensity of the beam.

Therefore

$$\frac{\partial I}{\partial x} = -I\rho.$$

If we let $R(x, y, z) = \log I(x, y, z)$ we have

$$\frac{\partial R}{\partial x} = -\rho$$

and

$$R(x_0, y, z) - R(x, y, z) = \int_{x_0}^{x} \rho(x, y, z)dx.$$

For convenience, we drop the z coordinate.

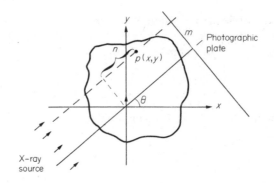

FIG. 6.2.

Figure 6.2 shows a parallel beam now inclined at θ to the x-axis, penetrating a body of finite extent, and falling on a photographic plate which, since we have dropped the z coordinate, is to be thought of as one-dimensional. If the initial log intensity is R_0 and if the log intensity at position m on the plate is given by $R_{m,\theta}$, then from the preceding analysis,

$$f(m, \theta) = R_0 - R_{m,\theta}$$

$$= \int_{-\infty}^{\infty} \rho(n\cos\theta - m\sin\theta, \quad n\sin\theta + m\cos\theta)\,dn$$

We can take an infinite integral because we have assumed that the body is of finite extent, i.e., ρ vanishes outside some region. The function f, as defined, can easily be computed from our knowledge of the initial intensity and from the optical density of the developed plate.

Now we compute the Fourier transform of f:

$$\Phi(\mu, \theta) = \int_{-\infty}^{\infty} f(m, \theta)\exp[-2i\pi m\mu]\,dm$$

$$= \int_{-\infty}^{\infty}\int_{-\infty}^{\infty} \rho(x, y)\exp[-2i\pi(x\mu\cos\theta + y\mu\sin\theta)]\,dx\,dy.$$

The importance of this last expression may not be immediately apparent. It is the two-dimensional Fourier transform of the density ρ, measured at position $(\mu\cos\theta, \mu\sin\theta)$ of Fourier transform space. Since it is possible to invert Fourier transforms, we can compute the original X-ray density, if we can measure f for all values of m and θ. It might be thought that there are less sophisticated methods of computing the density; and indeed there are other methods involving convolution integrals. However, the Fourier transform method is of practical importance, as Fourier transforms can be very rapidly computed by digital methods. In practice one would make discrete approximations to the various functions. It would also be more efficient to use an array of X-ray sensors rather than a large number of photographic plates.

*6.5.   Muscular Control

We now give some simple examples of muscular control.

Example 1. Muscular Movement

Although the man-machine analogy has not yet enabled us to predict or explain human behavior, it has helped in understanding the working of the body.

Nubar and Contini consider the dynamics of the human body in motion and at rest to derive a model of the body based upon theoretical mechanics. Considering the movement of the body in two dimensions, they apply the classic equations of dynamics (Newton's laws) to the segments of the body to obtain a system containing the following time-dependent unknowns:

(a) the coordinates of some definite point in the body, relative to some fixed axes, to which all other points may be referred;
(b) the orientations (angles) of all body segments, relative to the fixed axes,
(c) the moments at the ends of the segments at the joints; and
(d) reactions at all support points, but one.

It is assumed that the following quantities are given:

(e) the applied forces (weights);
(f) the physical characteristics of the segments (length, mass, center of gravity);
(g) the initial values of the unknowns and their first derivatives.

Since the moment at a joint is proportional to the muscle tension, we can define muscular effort as a function of the product of a joint moment and its duration. The simplest expression is $CM\Delta t$, where M is the moment at the joint, Δt the time duration, and C a numerical factor. Because muscular effort may be negative (in which case the joint moment is negative), the expression $CM^2\Delta t$ is adopted as a measure of muscular effort at a joint.

Assume that an individual will move (or adjust his position) in such a way as to reduce his total muscular effort to a minimum, consistent with the constraints. Thus, he seeks to minimize the effort

$$E = (c_1 M_1^2 + c_2 M_2^2 + \ldots + c_m M_m^2)\Delta t + A_0,$$

subject to the equations of motion.

To illustrate the operation of the principle of minimum effort in removing the indeterminacy of the equations, let us consider a simple example of a human being standing at rest with one foot on the ground (Fig. 6.3). It consists of five rigid parts: arms, legs, and trunk. The physical characteristics of the segments (length l_1, mass m_1, moment of inertia about the center of gravity I_1, moment of inertia about one end I_{1a}) are known. The location of the center of gravity of every segment is also known to be at a distance $k_i l_i$ from a joint, where k_i is a numerical factor. The system, fixed at point A, leaves the primary unknowns as the five orientations, $\alpha_1 \ldots \alpha_5$ with respect to the vertical and the four independent joint moments, M_{1B}, M_{2B}, M_{4D}, M_{5D}. These nine unknowns are underlined in the diagrams.

The simplest forms of the moment equations are found by considering that every segment is in equilibrium under the effect of inertial forces and inertial moments, plus the forces and moments applied at its end as reactions (d'Alembert principle). Minimization of

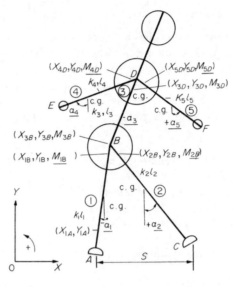

FIG. 6.3.

E with respect to the equations of motion produces the following solution:

$$M_{1B} = \frac{l_1}{A_1} \frac{s_0}{q^2}; \quad M_{2B} = \frac{l_2}{A_2} \frac{s_0}{q_2}; \quad M_{4D} = 0; \quad M_{5D} = 0,$$

$$\sin \alpha_1 = -\frac{l_1}{A_1^2} \frac{s_0}{q^2}; \quad \sin \alpha_2 = \frac{l_2}{A_2^2} \frac{s_0}{q^2},$$

$$\sin \alpha_3 = \frac{1}{A_3}\left(\frac{l_1}{A_1} + \frac{l_2}{A_2}\right); \quad \alpha_4 = 0; \quad \alpha_5 = 0,$$

where

$$q^2 = (l_1/A_1)^2 + (l_2/A_2)^2$$

and

$$A_1 = (K_1 m_1 + m_2 + m_3 + m_4 + m_5)l_1 g,$$
$$A_2 = m_2 K_2 l_2 g,$$
$$A_3 = (K_3 m_3 + m_4 + m_5)l_3 g,$$

and s_0 is a constant. For the case of an individual 5 ft 9 in, in height, weighing 160 lb, the solution is shown in Fig. 6.4.

*Example 2. Optimal Gaits

There are also models to determine optimal step-size or gait. We present two models, the first by Smith who determines the optimal gait of an animal (human) traveling fast on level ground and the second by Rashevsky who studies the optimal step-sizes for a human walking on level ground and uphill.

Smith states that when traveling fast (running, galloping, trotting), an animal (human)

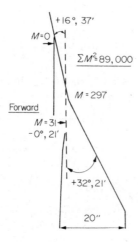

FIG. 6.4.

spends part of the time with all legs off the ground and part with one or more legs on the ground. The path of an animal's center of gravity is shown in Fig. 6.5.
It has been verified experimentally that

$$d \approx h\frac{a}{b}$$

(where d, h, and b are as shown in figure) which would be exact if the upward acceleration were uniform during the stepping phase.

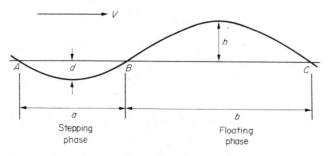

FIG. 6.5.

The criterion chosen to determine the optimal gait is the minimization of the power output. The total work done in a single stride is

$$\bar{W} = W_f + W_s,$$

where W_f is the work done in raising the center of gravity and W_s is the work done in accelerating the legs.

The time which elapsed between leaving the point B and the highest point of the floating phase is $(b/2V)$, where V is the velocity and, hence

$$h = \tfrac{1}{2}gt^2 = \frac{g b^2}{8 V^2}.$$

Thus

$$W_f = (h + d)mg = \frac{(a + b)bmg^2}{8V^2}.$$

The work done in accelerating the legs can be determined by replacing the mass of the whole limb by an equivalent mass m' at the foot

$$W_s = \tfrac{1}{2}m' V^2.$$

Hence,

$$\overline{W} = \frac{(a + b)bmg^2}{8V^2} + \tfrac{1}{2}m' V^2$$

and the work done per unit time (since the time to complete a stride is $a + b/V$) is given by $\overline{P}$, where

$$\overline{P} = \frac{\overline{W}V}{a + b} = \frac{bmg^2}{8V} + \frac{m' V^3}{2(a + b)}.$$

Now, if b is replaced by ja, where j can be regarded as a measure of the jumpiness of the gait and L represents a linear dimension of the animal in question (e.g., height, length of a limb), then m, m', are proportional to L^3, a is proportional to L.

Consequently

$$\overline{P} = C_1 j \frac{L^4}{V} + C_2 \frac{V^3 L^2}{(1 + j)}$$

and

$$\frac{d\overline{P}}{dj} = C_1 \frac{L^4}{V} - C_2 \frac{V^3 L^2}{(1 + j)^2} = 0,$$

where

$$(1 + j)^2 = \frac{C_1}{C_2} \frac{V^4}{L^2} \propto \frac{V^4}{L^2}.$$

Thus, $1 + j$ is proportional to V^2/L and hence j increases with V and decreases with L.

Some representative results are $j = 0$ for an elephant, $j = 0.3$ for a horse, and $j = 1$ for a greyhound. However, j does not always change as rapidly with L as the above examples suggest: it seems that there are additional criteria determining the optimum besides the minimization of power output.

Rashevsky considers a slightly different approach. He first defines the following variables:

$s = $ the length of the step,
$n = $ the number of steps per unit time,
$v = $ velocity of walking defined by $n s$,
$\Delta = $ the distance that the center of gravity is lifted,
$M = $ mass of the body,
$g = $ acceleration constant.

The power loss in walking n steps is then given by

$$W_\Delta = n \, \Delta \, Mg = \frac{Mgv \, \Delta}{s}.$$

He argues that the quantity Δ is determined by a combination of the length of the step and the length of the legs. Considering the legs as rigid, the difference Δ between the highest and the lowest positions of the center of gravity is

$$\Delta = l(1 - \cos \theta),$$

where l is the length of the legs and θ is angle between position of leg and vertical. However, if the position of the legs is approximated as shown in Fig. 6.6, then

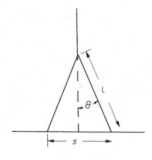

Fig. 6.6.

$$\sin \theta = s/2l$$

giving

$$\Delta = l(1 - \sqrt{1 - s^2/4l^2})$$

and

$$W_\Delta = Mgv \frac{l}{s}(1 - \sqrt{1 - s^2/4l^2}).$$

For steps which are not too large, $s/l < \frac{1}{2}$ and $s^2/4l^2 < 1/16$. So we may approximate

$$\left(1 - \frac{s^2}{4l^2}\right)^{1/2} \quad \text{as} \quad \left(1 - \frac{s^2}{8l^2}\right) \text{ and}$$

thus

$$W_\Delta = \frac{1}{8} Mgv \frac{s}{l}.$$

The power loss due to the imparting of kinetic energy to the swinging extremity is found to be

$$W_e = \frac{1}{2} I \frac{v^3}{sl^2},$$

where I is the moment of inertia of the extremity with respect to the hip joint.

Thus, total power loss is

$$W_t = \frac{Mgvs}{8l} + \frac{mv^3}{8s},$$

where m is the mass of the limb.

The criterion chosen to determine the optimal step-size s_0 is to minimize the total power loss. The result is

$$s_0 = v\sqrt{\frac{ml}{Mg}},$$

which depends upon v.

Example 3. Best Velocity in a Running Race (control theory)

How does an athlete run a race? That is, how does he husband his efforts in order to cover the distance D in the shortest possible time? Intuitively, it is clear that the athlete's optimal strategy is dependent upon D. When D is small, the best course is to run "flat out" since the runner will not exhaust all of his energies. If D is large, on the other hand, he must "pace" himself so that by the end of the race, but not before, he will have expended all of his strength.

Keller shows one way to model this highly practical physical situation. We want to advise the runner how to vary his speed $v(t)$ during a race of distance D in order to minimize the time T required. D, T, and v are related by the equation

$$D = \int_0^T v(t)dt. \tag{6.2}$$

The velocity v satisfies

$$\frac{dv}{dt} + \frac{v}{\tau} = f(t), \tag{6.3}$$

where v/τ is the resistance per unit mass (τ is a given constant) and $f(t)$ is the thrust per unit mass.

Initially

$$v(0) = 0. \tag{6.4}$$

The force $f(t)$ is controlled by the runner but cannot exceed a constant F, so we have

$$f(t) < F. \tag{6.5}$$

If $E(t)$ denotes the mass of oxygen (per unit mass) available to the athlete's muscles, then since oxygen is metabolized by the body to produce energy and energy is used at the rate fv by the athlete, we have

$$\frac{dE}{dt} = \sigma - fv, \tag{6.6}$$

where σ is the rate at which oxygen is supplied by breathing and circulation. Initially,

$$E(0) = E_0, \tag{6.7}$$

and, since $E(t)$ is never negative,

$$E(t) > 0. \tag{6.8}$$

The athlete's problem can now be described as follows: find $v(t)$, $f(t)$, and $E(t)$ satisfying (6.3) through (6.8) so that T, defined by (6.2) is minimized. The physiological parameters τ, F, σ, and E_0 and the distance D are specified in advance.

We combine (6.3) and (6.5) to give

$$\frac{dv}{dt} + \frac{v}{\tau} \leqslant F. \tag{6.9}$$

We can also use (6.3) to eliminate f from (6.6). Integrating the resultant equation and using the initial condition (6.7) yields

$$E(t) = E_0 + \sigma t - \frac{v^2(t)}{2} - \frac{1}{\tau}\int_0^t v^2(s)ds. \tag{6.10}$$

Combining this with (6.8) gives

$$E_0 + \sigma t - \frac{v^2(t)}{2} - \frac{1}{\tau}\int_0^t v^2(s)ds \geqslant 0. \tag{6.11}$$

Since we have expressed f and E in terms of v, our original problem can be expressed as follows: Find $v(t)$ satisfying (6.4), (6.9), and (6.11) so that T is minimized.

Now minimizing T subject to fixed D is equivalent to maximizing D with T given, which is the formulation we shall consider.

To begin with, we can assume $f(0) = F$. (The rate fv of doing work is 0 initially since $v(0) = 0$ and so $f(0)$ can be taken as large as possible.) Hence, $f(t) = F$ for $0 < t < t_1$, where $0 < t_1 < T$ and t_1 is as large as possible.

It is clear intuitively (and confirmed by analysis) that for T not too large, $t_1 = T$. The critical value for T is shown to be $T = T_c$, where T_c is the unique positive root of the equation

$$E_0 + \sigma t - F^2\tau^2\left(\frac{t}{\tau} + e^{-t/\tau} - 1\right) = 0.$$

If $T > T_c$, then we must consider the "other" end of the interval $0 < t < T$. Just as we assumed $f(0) = F$, we can also assume $E(T) = 0$ (for otherwise he could have run harder). Thus, $E(t) = 0$ for $t_2 < t < T$, where $t_1 < t_2 < T$ and t_2 is as small as possible.

It is very interesting to note that when D is very much larger than D, where D_c is the distance corresponding to T_c, then we may assume $t_1 = 0$ and $t_2 = T$. One can verify the validity of this assumption analytically.

On a heuristic basis, this just means that one should try to run steadily over a long distance.

Using the preceding formulae and the results of actual races, Keller has calculated the values of τ and F and, using them, the values of σ and E_0. D_c is shown to be 291 m, a not unreasonable value. In Fig. 6.7 the curve represents the calculated value for average velocity D/T for $D < 2000$ m; the points are derived from world records.

A final caveat is in order here. From the fact that a model appears to fit the observed data, one should never allow oneself to conclude that the model must be correct. Only a failure to meet nature gives one the right to draw a conclusion and then it must be to reject

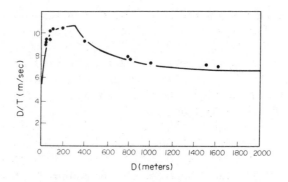

FIG. 6.7.

the model. While Keller's model fits the observed data quite well, his assumptions do not include a direct dependence of F on $E(t)$, i.e., on t. Yet one can certainly argue that the maximum F available to an athlete does depend on the time.

In another paper dealing with the mathematics of athletics, Brearley has shown by a careful mathematical analysis that the sudden increment of over 2.5 ft in the new world's record for the long jump established at the 1968 Olympic Games in Mexico City could not be attributed to the decreased air resistance one encounters at higher altitudes, since an upper bound to the possible increment would have been 2.5 in.

6.6. Weight Control

One area which draws on both physics and biology is weight control.

Example. Weight Control and Energy (Lloyd P. Smith)

Body weight is determined by the energy taken in and the energy ejected from the body as heat. The energy E_i taken in over a 24-hour period is transformed to other forms, stored as fat, or ejected as heat. The law of conservation of energy gives for ΔE the energy stored per day: $\Delta E = E_i - I - E_0$, where I is the energy required to sustain life and E_0 is the average energy which leaves the body during a 24-hour period. A pound of weight is added whenever $\Delta E = 3000$ Kcals (kilogram calories, which are popularly called calories. It is the amount of heat required to raise one kilogram of water one degree centigrade or 1.8 degrees Fahrenheit. It takes 3086 ft-lb of exercise to equal one kilocalorie of body energy.) If the ideal weight in pounds is W_0 and the weight is W, we have $W - W_0 = \Delta E/3000$. Now for each pound of weight the body needs an additional 14 Kcals per day to maintain it at body temperature and carry it around. Thus the daily rate of change of body weight is

$$3000 \frac{dW}{dt} = \Delta E - 14(W - W_0)$$

whose solution is given (in pounds) by:

$$W - W_0 = \frac{\Delta E}{14}[1 - e^{-14t/3000}].$$

If we assume that ΔE is constant from day to day, the equilibrium body weight is obtained by putting $dW/dt = 0$ as the weight W will no longer change with time. We have $W - W_0 = \Delta E/14$. Thus if $\Delta E = \pm 200, \pm 400, \pm 600, \pm 800$ Kcals per day, $W - W_0 = 14.3, 28.6, 42.8, 57.1$ lb gained or lost. The time required to gain or lose one pound is 15.5, 7.6, 5.1, and 3.8 days.

Now we study the factors which determine E_i, I, and E_0. For a 25-year-old man with 150 lb weight, sleeping or lying down, surrounded by air of 68°F, on the average, I is 70 Kcals per hour. For a woman weighing 128 lb, I is 62 Kcals per hour. This would be the absolute minimum for E_i. If we assume that I is a constant for a given individual we are left with $E_i - E_0$, which affects ΔE, to control; gain if positive, loss if negative. Now E_i could arise from the following kinds of gain: (1) energy in chemical form (food or drink) through the stomach (the most important); (2) heat energy absorbed from surroundings; (3) heat from hot food or drink; (4) sunbathing; (5) energy from mechanical work on body, such as massage.

1. Chemical energy is converted to heat by the liver and involuntary muscles, the latter through tensing, relaxing, standing, shivering, or responding to emotion and distributed to the body through the blood stream and not by conduction since the core temperature of the body, 98.6°F, is the same everywhere. The caloric content of most foods is known in the literature. Conversion processes are controlled by large protein molecules called enzymes maintaining the internal environment of the body cells as constant as possible. Enzyme deficiency may require a higher value for E_i than I to maintain basic body energy needs. Different people's food consumption depends on the efficiency of their enzymes. An enzyme called Steapsin decreases the size of normal fat particles so they can pass through the intestinal wall.

2. Unless the heat surrounding the body is very high, the body does not absorb heat. High temperature tends to prevent the body from getting rid of heat.

3. Heat would be distributed to the body if the drink is warmer than 98.6° F and taken from the body in the opposite case to raise the temperature of the liquid.

4. Although the contributions of sources under 2 and 3 are negligible, not so with solar energy. There are 14.3 Kcals per minute incident on 1 square meter of surface. The human body has 1.55 square meters; if we use half of it we obtain 11.17 Kcals per minute or 670 Kcals per hour. Although some would be reflected, 70% will be absorbed, i.e., 469 Kcals per hour, which is high when compared with the 70 Kcals per hour needed for basic energy requirements. It must first be ejected before calling upon stored energy as fat. The figures here may seem high, but even if we take 40% of these values, they indicate that obese people might well be advised to stay out of the sun.

5. The direct result of massage, which requires considerable mechanical energy, is negligible except for breaking down fat particles to make it easier for the body to oxidize them.

Decrease in weight is due to energy loss. There is loss from:

1. Increasing the temperature of ingested food, e.g., a 12-ounce glass of ice water requires 13 Kcals to bring it up to body temperature, or a reduction of 0.0043 lb. It would take 233 glasses of such water to lose one pound.

2. Loss of moisture exhaled at body temperature—a less-efficient mechanism than in 1.

3. Elimination of liquid and solid waste. Difficult to estimate because of the varying enzymatic activities of different people.

4. Energy required to stand or sit. Sewing or clerical work require 24 Kcals per hour, standing around, 92 Kcals per hour. The loss is not significant.

5. Doing significant mechanical work such as exercising. Consider a 150-lb man walking at 2 miles per hour. Assume that the body is raised 1 in. on each stride of 2 ft length. The arithmetic gives $\frac{2 \times 5280}{2} \times \frac{150}{12} = 66{,}000$ ft-lb or 21.4 Kcals. The measured value of energy expended for this purpose is 115 Kcals. Thus the body generates 93.6 Kcals of heat to perform 21.4 Kcals of mechanical work. If the same man runs at 8 miles per hour, with a 2-ft stride and raising the body 2 in., the result is 171 Kcals per hour, but the measured energy expenditure is 730 Kcals of heat, i.e., 559 Kcals of heat to perform 171 Kcals of mechanical work. The weight loss is 0.0072 lb in the first case and 0.057 lb in the second for the mechanical work and 0.038 lb and 0.243 lb, respectively, from heat loss—5.3 and 4.3 times the mechanical work done. The lungs of a person at rest draw in and expel $500 \, \text{cm}^3$ per breath. Strenuous exercise raises this figure by tenfold and increases the breathing rate—with considerable heat loss.

6. Loss of heat through body surface—the most important means of energy loss. The layer just under the skin consists of adipose tissue (thicker in women) storing fat globules which serves as an insulator. Fat deposits are being constantly used and reformed. Fat is stored in tissues least influenced by muscular activity: the abdominal region, the waist, neck, and the buttocks. (The hands act as an excellent medium of heat exchange—they get hot or cold as necessary.) If the skin temperature drops below 91.4°F or if the body temperature drops below 98.6°F, the blood vessels in the surface fat contract, decreasing heat flow through the fat layers to the surface and heat is generated in the body by conversion of food or stored fat. Thus the surface temperature must be as near as possible to body core temperature to stimulate evaporative cooling by the presence of water on the body surface. We estimate heat and weight losses by heat conductivity and by evaporative cooling.

First let us estimate the heat loss by heat conductivity when the blood circulation keeps the temperature of the adipose tissue 0.05 cm below the body surface at the body temperature of 37°C and the skin temperature T_s°C. The rate of heat loss would be

$$\dot{H} = \frac{3.6K(37 - T_s)}{0.05} \; \text{Kcals/cm}^2 \text{ per hour,}$$

where K is the heat conductivity of the thin layer of adipose tissue which is taken as $0.00044 \, \text{cal/cm}^2$ per °C. Taking the skin temperature as 20°C or 68°F,

$$\dot{H} = 0.54 \; \text{Kcals/cm}^2 \text{ per hour.}$$

If the blood vessels in the adipose tissue were completely contracted and the adipose tissue is 2 cm thick,

$$\dot{H} = 0.014 \; \text{Kcals/cm}^2 \text{ per hour,}$$

which indicates the kind of control that the autonomic nervous system can exert on the magnitude of heat loss.

As will be indicated in the following calculation, the largest amount of heat loss can be effected by evaporative cooling. To estimate this we use the formula giving for the number of molecules of water evaporating per cm^2 per sec:

$$N_e = v\, Rn \exp \frac{-W}{kT},$$

where N_e is the number of water molecules that will evaporate per sec per cm^2 when there is a layer of water (perspiration) on the skin at an absolute temperature T in $°K$ which we will take as the body temperature of $37°C$ or $310°K$. v is the average velocity of a water molecule in the liquid at temperature $310°K$ and is given by $1/2\, Mv^2 = 3/2kT$. n is the number of water molecules per cm^3 of liquid, R is the ratio of the number of molecules hitting the surface of the water from the vapor phase that recondense to those that are reflected. For our purpose, $R = 1$, k is Boltzman's constant $= 1372 \times 10^{-16}$ ergs/degree, and v is the energy in ergs that must be expended in removing a water molecule from the liquid state to the vapor state at a temperature of $37°C$. $v = 725 \times 10^{-15}$ ergs. Using these values, $N_e = 3.13 \times 10^{23}$ molecules/cm^2 per hour $= 9.35$ g of water per hour per cm^2. This can be translated into the number of Kcals removed from an area of 1 cm^2 per hour from the body by evaporative cooling. It is $H_e = 5.45$ Kcals/cm^2 per hour. Comparing this figure with 0.54 given in conductive heat loss we see that the heat loss by evaporative cooling can be about ten times more effective than the loss by pure heat conductivity.

From these principles we learn that to lose weight E_i must be less than $I + E_0$. Since E_i is mainly determined by the caloric value of ingested food and the absorption of the sun's radiation (sun bathing), these should be controlled. To make E_0 as high as possible an effort should be made to keep the skin temperature as close as possible to the body core temperature and the outside temperature as low as can be reasonably managed.

From tables of E_i and E_0 it can be concluded that the value of ΔE can be much more influenced by one's exercise program than the value of E_i unless one insists on consuming large quantities of high calorie foods like fat meats, pies, and rich cakes, roast chicken, cheeseburger, pizza, and recreational beverages.

With this method of weight control, one knows precisely the factors to adjust and how effective each will be in controlling weight with no pills, some of which could have harmful side effects, and are not necessary nor desirable. Also, one's body will increase in physical fitness and be capable of increased accomplishment.

6.7. Cellular and Genetic Applications

We give some examples from the fields of cell biology and genetics. The next two examples show how a simple mathematical analysis of genotypes may be used to study how traits are distributed in a population.

Example 1. The Hardy–Weinberg Law of Equilibrium

Let A and a be dominant and recessive genes, respectively, controlling some physiological trait. Suppose that A occurs with probability p in some population while a occurs with probability $q = 1 - p$. Since every individual inherits two genes for each (nonsex-linked) trait, the possible genotypes are AA, Aa, and aa. The Hardy–Weinberg formula, discovered independently in 1908 by G. H. Hardy, a British mathematician, and

W. Weinberg, a German physician, states that in a large population where mating is random—at least with regard to the trait controlled by A and a—the distribution of genotypes after one generation is as follows: AA occurs with probability p^2, Aa with probability $2pq$, and aa with probability q^2.

It is easy to derive this law from the elementary theory of probability. Let $P(E)$ denote the probability of some event E. If F_A denotes the event that some individuals' father contributes A and M_A is the corresponding event for his mother, then F_A and M_A are independent events since mating is assumed random with regard to A. Hence, $P(F_A$ and $M_A) = P(F_A)P(M_A) = pp = p^2$. Thus, the probability that some individual has genotype AA is p^2. Similarly, the probability of genotype aa is q^2. These genotypes, in which both genes are the same, are called *homozygous*. A genotype in which both genes are different is called *heterozygous*. If m denotes the probability that a genotype is homozygous and t the probability that it is heterozygous, then $m + t = 1$; that is, $t = 1 - m$. Now $m = p^2 + q^2$, so $t = 1 - p^2 - q^2$. Finally, note that $1 = 1^2 = (p + q)^2 = p^2 + 2pq + q^2$ and, therefore, $t = 1 - p^2 - q^2 = 2pq$. But t is just the probability that Aa occurs.

This may also be seen by noting that Aa can arise in two ways, with A inherited from the father and a from the mother, and vice versa.

Note that, as usual, the Hardy–Weinberg model gives a necessary condition—in this case, for random mating. Observe also that mating may be random with regard to some factors (e.g., blood group) but highly selective with respect to others (for example, skin pigmentation). Hardy–Weinberg equilibrium can be used, in conjunction with gene frequency analyses, as a test for nonrandom mating; that is, if the distribution of genotypes differs significantly from their predicted frequencies, then mating is not random with respect to the factor under consideration.

Example 2. Blood Groups

Following the last example, we can assume that the blood group of an individual is determined by the genes of the two parents. The genes that determine the blood group are of three types, G^A, G^B, and g. G^A and G^B are dominant genes and g is the recessive gene. The blood groups O, A, B, and AB are determined as follows:

$$g \times g \to 0, \quad G^A \times g \to A, \quad G^B \times g \to B, \quad G^A \times G^B \to AB$$
$$G^A \times G^A \to A, \quad G^B \times G^B \to B.$$

What is the proportion of genes G^A, G^B, and g in the parent population?

p is the probability that an individual has gene G^A.
q is the probability that an individual has gene G^B.
$1 - p - q = r$ is the probability that an individual has gene g.

Assuming that individuals mate randomly, we can find the probability that the offspring has a particular blood group.

$$\pi_1 = P(O) = P(g, g) = r^2,$$
$$\pi_2 = P(A) = P(g, G^A) + P(G^A, g) + P(G^A, G^A) = rp + pr + p^2 = p^2 + 2pr,$$
$$\pi_3 = P(B) = P(g, G^B) + P(G^B, g) + P(G^B, G^B) = rq + qr + q^2 = q^2 + 2qr,$$
$$\pi_4 = P(AB) = P(G^A, G^B) + P(G^B, G^A) = pq + qp = 2pq.$$

We need to find the "best" estimates of p, q, r. One way of defining "best" is that p^*, q^*, r^* are the best estimates from a sample S if $P(S|p^*, q^*, r^*) \geqslant P(S|p, q, r)$ for all p, q, r such that $p + q + r = 1$. Such an estimate is termed the maximum likelihood estimate.

Now, if n is the sample size and n_1 have blood group O, n_2 have group A, n_3 have group B, and n_4 have group AB, where $n = n_1 + n_2 + n_3 + n_4$, then L, the probability that this arises given any p, q, r, is

$$L = P(\text{sample}|p, q, r) = \frac{n!}{n_1! n_2! n_3! n_4!} \pi_1^{n_1} \pi_2^{n_2} \pi_3^{n_3} \pi_4^{n_4}.$$

We wish to maximize this subject to $p + q + r = 1$, i.e., using Lagrange multipliers. The mathematics is cumbersome and we omit the solution.

We now look at an example in chromosome mapping.

*Example 3. Renewal Theory and Chromosome Mapping [Bailey]

Renewal processes can be used to explain the phenomenon of genetic linkage and chromosome mapping. The points of exchange which occur on a single chromosome strand during the appropriate stage of meiosis can be regarded to follow the renewal process. The concept of breakdowns up to time t is replaced by the points of exchange that occur along the strand, which is represented by a semi-infinite straight line with origin corresponding to the chromosome's centromere. Assuming that the intervals between successive points of exchange are independently distributed with identical frequency functions $f(u)$, the Laplace transform $X^*(s)$ corresponding to $X(t)$, which is the average number of points of exchange in the interval $(0, t)$ is given by

$$X^*(s) = \frac{f^*(s)}{s\{1 - f^*(s)\}},$$

where

$$f^*(s) = \int_0^\infty e^{-st} f(t) dt,$$

is the Laplace transform of f.

By putting $f(u) = e^{-u}$ we obtain

$$f^*(s) = \frac{1}{1+s}$$

and

$$X^*(s) = 1/s^2.$$

Thus
$$X(t) = t.$$

The recombination fraction, which is the probability of an odd number of points of exchange in the interval $(0, t)$ is given by

$$y(t) = \sum_{n=0}^\infty p_{2n+1}(t) = \tfrac{1}{2}\{1 - p(-1, t)\}$$

and its transform is given by

$$y^*(s) = \tfrac{1}{2}\{s^{-1} - p^*(-1, s)\} = \frac{f^*(s)}{s\{1 + f^*(s)\}}.$$

Substituting $f^*(s) = 1/(1+s)$ we get $y^*(s) = 1/2(1 - e^{-2t})$, which is known as Haldane's formula.

Our next example considers the control of protein synthesis.

Example 4. Protein Synthesis [J. M. Smith]

We discuss methods taken from the kinetics of chemical reactions used to analyze the control of protein synthesis. The simplest model for the control of protein is as shown in Fig. 6.8.

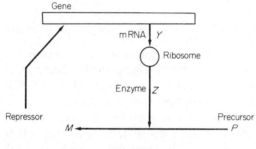

FIG. 6.8.

The mRNA is made in the nucleus, its concentration at any moment being Y. At the ribosomes the "message" is translated, and enzyme molecules synthesized, their concentration at any moment being Z. This enzyme catalyzes the reaction from an inactive precursor, concentration P, to a repressor molecule, concentration M. The repressor molecule then reacts with the gene, so that when a repressor is attached to a gene no mRNA is made.

If the rate at which mRNA molecules are lost or destroyed is assumed proportional to their concentration, then

$$\frac{dY}{dt} = \frac{c}{a + bM} - kY,$$

where a is the (constant) probability that any particular repressor molecule will become detached in a given time interval, and b, c, and k are constants.

Similarly,

$$\frac{dZ}{dt} = eY - fZ,$$

where e and f are constants, and

$$\frac{dM}{dt} = gZ - kM,$$

where g and h are constants.

These equations can be simplified if one is interested only in Y and Z, since the precursor–repressor reaction will reach equilibrium much more rapidly than the others, i.e., $dM/dt = 0$. Thus, writing $bg/h = 1$,

$$\frac{dY}{dt} = \frac{c}{a+z} - kY,$$

$$\frac{dZ}{dt} = eY - fZ.$$

The important question is whether these equations indicate a sustained oscillation or whether any disturbance is rapidly damped out. The equations can be solved to show that they describe either a damped oscillation or no oscillation, but if a slight modification is made to account for the time lapse while mRNA molecules travel from gene to the ribosome, then the control system is oscillatory.

*6.8.   Models related to the Nervous System

Because of its highly complex nature and the difficulty of studying it directly, the nervous system makes an excellent subject for mathematical modeling. A leading authority in this field is Nicholas Rashevsky.

Example. The Central Nervous System [Rashevsky]

Neurons are connected by *synapses* in *nets*. Nerve conduction is unidirectional for any neuron; it is chemical in nature and follows the laws of Boolean algebra. Functionally, a *pathway* is a group of axons between the brain and an organ in an "anatomically discernible tract." Pathways from a sense organ to the brain are *afferent*; those from the brain to a motor functionary are *efferent*. For a weak stimulus, the excitation energy E along a pathway is assumed to be approximately linear:

$$E = \alpha(S - h), \tag{6.12}$$

where S is the stimulus intensity and h is the threshold of the "weakest link" in the pathway.

This is only an approximation since E cannot increase indefinitely with S. A better approximation is given by

$$E = \alpha h \ln \left\{ \frac{S/h}{1 + \dfrac{\delta S}{h}} \right\} \tag{6.13}$$

where

$$\delta = \exp[-E^*/\alpha h]$$

and

$$E^* = \lim_{S \to \infty} E.$$

Some neurons form closed cycles connecting pathways as shown in Fig. 6.9.

Once stimulated, the cycle continues indefinitely (reverberates) until chemical failure occurs at random. If ε is the number of excited cycles in a given pathway, then the failure

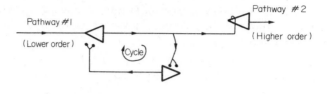

FIG. 6.9.

rate may be expressed by

$$\frac{d\varepsilon}{dt} = AE - a\varepsilon \tag{6.14}$$

since it has a rate of increase proportional to E and a rate of decrease proportional to ε.

Cycles may also inhibit pathways on which they synapse. If j inhibitory cycles are excited, then

$$\frac{dj}{dt} = BE - b_i, \tag{6.15}$$

where E is the intensity of the signal at the higher-order pathway.

If $E > j$, then the higher-order pathway exhibits a net excitation, and vice versa. The solution to (6.14) is

$$\varepsilon = \frac{AE}{a}(1 - e^{-at}) \tag{6.16}$$

plotted in Fig. 6.10, together with the extinction curve when the signal E is removed at time t_1.

The extinction equation (for E removed) is seen to be

$$E = \frac{A}{a}e^{-a(t-t_1)}. \tag{6.17}$$

Similar results apply to j.

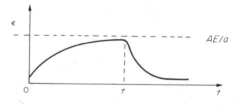

FIG. 6.10.

Now consider a simple reflex arc of purely excitatory pathways, one afferent (a.p.) and one efferent (e.p.) joined together. Let a stimulus of intensity S be an input to the a.p. resulting in an intensity E given, say, by $E = \alpha(S - h_1)$, with h_1 as the threshold of the a.p. Let τ_A be the conduction time of the a.p. Then, if h_2 is the threshold of the e.p., the efferent excitation begins with $\varepsilon = h_2$ for afferent cycles. We combine this with (6.16) to find t_0, the time at which the e.p. becomes excited.

$$\frac{AE}{a}(1 - e^{-at_0}) = h_2 \tag{6.18}$$

or

$$t_0 = \frac{1}{a} \ln\left(\frac{AE}{AE - ah_2}\right). \tag{6.19}$$

The total reaction time is thus

$$\tau_R = \tau_A + \tau_E + \frac{1}{a} \ln\left(\frac{AE}{AE - ah_2}\right), \tag{6.20}$$

where τ_E is the conduction time along the e.p. Substituting for E from (6.12), we obtain

$$\tau_R = \tau_A + \tau_E + \frac{1}{a} \ln\left(\frac{\alpha A(S - h_1)}{\alpha A(S - h_1) - ah_2}\right). \tag{6.21}$$

We see that τ_R is imaginary for $\alpha A(S - h) < ah_2$ (i.e., for $AE/a < h_2$) since the asymptotic value of ε is still less than the threshold of the e.p. For $\alpha A(S - h_1) = ah_2$ the reaction time is infinite; otherwise the time decreases with increasing S. [For large S, (6.21) no longer applies, as E is more closely approximated by (6.13) than by (6.12).] Experimental data have been obtained which closely fit the theoretical results obtained from (6.21).

6.9. Evolution

The method used to analyze the process of evolution is taken from the field of kinetics. In physical chemistry we view the progressive changes in a system comprising several chemical elements by enumerating the components, stating their character, and deriving an expression for the instantaneous state of the system in terms of significant parameters; this expression for the instantaneous state of the system is then used to determine its history. For example, if hydrogen, oxygen, and steam are put in a container of volume v at a given pressure and temperature, then the change in the mass of the steam m_1 is given by

$$\frac{1}{v}\frac{dm_1}{dt} = k_1 \frac{m_2^2 m_3}{v^2} - k_2 \frac{m_1^2}{v^2},$$

where m_2, m_3 are the masses of the hydrogen and oxygen, respectively, and k_1, k_2 are coefficients reflecting the temperature, pressure, and the reaction. Our interest in this equation lies in its general form

$$\frac{dm_1}{dt} = F(m_1, m_2, m_3; v, T)$$

and it is this relation that is transplanted into the field of organic evolution.

The masses of the chemical components are replaced by the population of the species and the parameters in the equation represent the environmental conditions, e.g., climate, topography. In particular, one form of the general relationship is

$$\frac{dN_i}{dt} = e_i N_i + \sum_{s=1}^{n} a_{is} N_i N_s,$$

where N_i is the population of the ith species and a_{is} represents the interaction between the species.

For single species with a limited food supply, this equation becomes

$$\frac{dN}{dt} = eN - aN^2.$$

If we let $K = e/a$, we obtain

$$\frac{dN}{dt} = aN(K - N),$$

which shows that the population N increases geometrically when the species is very rare but gradually levels off toward the equilibrium value K, the maximum population it can maintain.

6.10. Social Biology

Social factors in biology consist of interspecies and intraspecies interactions, and these factors may often play an important role, e.g., infanticide in certain human societies. Predator–prey relations, in which one group depends upon another for its existence, are quite common and can easily be described mathematically.

We give a simple example.

Example. Predator–Prey Systems [Bailey]

The predator–prey relationship, similar to that discussed in Chapter 3 but without a memory term, is closely illustrated in the case of wolves which, for the most part, prey on the "surplus crop" of deer, moose, or caribou. The wolves, like other creatures, eat to sustain themselves. Since they use less energy to pursue a meal than they realize from it, wolves tend to attack the weak, old, or sickly. The general fitness of the deer or caribou herd is heightened, and its population is kept at a level for which there is an adequate food supply. In this way, predator and prey help each other. Should the deer population become thin, wolves must decrease in number, as in fact happens. Campaigns to exterminate the American red wolf to extinction levels created an opportunity for the coyotes, which feed on the same prey, to multiply with greater threat to ranchers than the wolves ever were.

Now for the analysis.

Let $N_1(t)$ be the population of the prey, and $N_2(t)$ of the predator, at time t. If N_2 is small, N_1 increases, entailing an increase in N_2. This increase in N_2 is followed by a decrease in N_1, which causes starvation and hence a decrease in N_2. The number of encounters of the two species is proportional to $N_1 N_2$. In an encounter one species decreases while the other increases. Thus

$$\frac{dN_1}{dt} = aN_1 - bN_1 N_2$$

and

$$\frac{dN_2}{dt} = -cN_2 + dN_1 N_2,$$

where a, b, c, d are all positive. If we divide the first equation by the second, integrate, and then substitute,

$$N_1 = x + \frac{c}{d} \equiv x + p,$$

$$N_2 = y + \frac{a}{b} \equiv y + q,$$

we obtain

$$c \log (x + p) + a \log (y + q) - dx - by = c,$$

where c is the constant of integration. Expanding series about the origin and neglecting terms of higher order, we obtain ellipses that describe periodic variations in the prey population (e.g., a host population) and in the predators (e.g., parasites). These ellipses are given by

$$\frac{cx^2}{p^2} + \frac{ay^2}{q^2} = D,$$

where D is a constant. The period of oscillation near the origin is given by $2\pi/(ac)^{1/2}$. In the next chapter we consider models in the social and behavioral fields.

Chapter 6—Problems

1. Can you formulate a model of the lung different from the one described in this chapter?
2. Evaluate the models of muscle control. Can you formulate others?
3. How would the equation for blood flow be modified if the radius of the "pipe" is very small (e.g. capillaries). Consider the effects on the walls.
4. A cyclist needs to know how to position himself in order to take into account the wind, the curvature of the path, etc. Formulate a simple model and derive some rules for him.
5. At what point is it physically possible for a baby to: (a) stand, (or b) walk? Formulate some criteria for this.
6. Formulate a model for mutations in cells, given that such mutation is an accidental fluctuation whose probability is a function of time. Suggest some appropriate forms for this function for given mutations, e.g., cancer cells.
7. Develop a model of a bionic limb. What are desirable characteristics? What information do you need? Use simple dynamic principles for your first model and then produce a more sophisticated model. (Use the information on muscular control.)
8. You have been asked to advise on stocking a river with fish. What information would you need about the surroundings and about the other animals in the area to help you decide on quantities of fish and the times for stocking? Derive a rough model for this. (Think about the predator–prey models.)
9. Consider a population in a given country. What determines the size of the population and the number in different age groups? What information would you need for estimating the number of people over 70?
10. How long would it take to die from starvation? What factors enter into this? [*Hint*: consider the weight-control problem.]

References

Bailey, Norman T. J., *The Elements of Stochastic Process with Applications to the Natural Sciences*, Wiley, New York, 1964.

Brearley, M. N., The long-jump miracle of Mexico city, *Math. Mag.*, Vol. 45, November 1972.

Collins, R. E., Kilpper, R. W., and Jenkins, D. E., A mathematical analysis of mechanical factors in the forced expiration, *Bull. Math. Biophys.*, Vol. 29, 1967, pp. 737–745.

Crowther, R. A., Derosier, D. J., and Klug, A., *Proc. Roy. Soc. Lond.* A, Vol. 317, 1970, p. 319.

Danziger, L. and Emergreen, G. L., mathematical models of endocrine systems, *Bull. Math. Biophys.*, Vol. 19, 1957, pp. 9–18.

Defares, J. G. and Sneddon, I. N., *An Introduction to the Mathematics of Medicine and Biology*, Year-Book Medical Publishers, Chicago, 1961.

Evans, J. W., Cantor, D. G., and Norman, J. R., the dead space in a compartmental living model, *Bull. Math. Biophys.* Vol. 29, 1967, pp. 711–718.

Keller, Joseph B., Optimal velocity in a race, *Am. Math. Monthly*, May 1974.

Nubar, Y. and Contini, R., A minimal principle in bio-mechanics, *Bull. Math. Biophys.* Vol. 23, 1961, pp. 377–391.

Rashevsky, N., *Mathematical Biology of Social Behavior*, University of Chicago Press, 1959.

Rashevsky, N., A note on energy expenditure in walking on level ground and uphill, *Bull. Math. Biophys.*, Vol. 24, 1962, pp. 217–227.

Smith, J. Maynard, *Mathematical Ideas in Biology*, Cambridge University Press, New York, 1968.

Smith, Lloyd P., *How to regulate your weight scientifically*, (published privately, 1980).

Spencer, R. P., A blood volume, heart weight relationship., *J. of Theoret. Biology*, Vol. 17, 1969, pp. 441–446.

Stacy, R. W., Barth, D. S., and Chilton, A. B., A mathematical analysis of oxygen respiration in man, *Bull. Math. Biophys.* Vol. 16, 1954, pp. 1–14.

Yilmaz, H., Psychophysics and pattern interaction, in Walthen–Dunn, W. (ed.), *Models for the Perception of Speech and Visual Form*, MIT Press, Cambridge, Massachusetts, 1967.

Bibliography

Barnoon, Shlomo, and Harvey Wolfe, *Measuring the Effectiveness of Medical Decisions: An Operation Research Approach*, Charles C. Thomas, Springfield, Illinois, 1972.

Cogan, F. J., R. Z. Norman, J. G. Kemeny, J. L. Snell, and G. L. Thompson, *Modern Mathematical Methods and Models*, Vol. II, Mathematical Association of America, 1958.

Defares, J. G. and I. N. Sneddon, *An Introduction to the Mathematics of Medicine and Biology*, Year Book Medical Publishers, Chicago, 1961.

Landany, S. P. and Machoi, R. E. (eds.), *Optimal Strategies in Sports* (Studies in Management Science and Systems, Vol. 5), Elsevier, Amsterdam, 1977.

Lotka, Alfred J., *Elements of Mathematical Biology*, Dover, New York, 1956.

Rashevsky, N., *Mathematical Biophysics*, Vol. I, Dover, New York, 1960.

Rashevsky, N., *Mathematical Principles in Biology and Their Applications*, Charles C. Thomas, Springfield, Illinois, 1961.

Chapter 7

Social and Behavioral Applications

7.1. Introduction

IN this chapter, we discuss modeling examples with the social and behavioral aspect as the leading theme; emphasis has been put on social–behavioral problems which are generally unstructured and difficult to model.

7.2. Courts and Justice

We consider measures of justice and injustice.

Example. Mathematical Model of Justice

This model is a first approximation to assigning values to levels of injustice due to crime and to waiting for trial in a community. It takes into account the types of crimes, the intensity felt in the community about each crime, the time lapse between apprehension and prosecution, the length of sentences (and the deviations from what they should be), and the possibilities for convicting an innocent man. Some uses of this model may be (i) the establishment of a comparative index to be evaluated for a community over a given time period and to be compared with other communities and time periods, and (ii) a second index to be used for the best organization of a docket: the "optimal" sequencing of trials.

Let $(\gamma_j)j = 1 \ldots n$ be the set of all types of crime. Each γ_j has a magnitude c_j (proportional to the sentence to be given), and a mean trial length l_j and sentence length m_j.

An individual crime i belonging to class τ may be assumed to have an average intensity $1/\beta_i$ depending on the character of the criminal and/or the victim and upon the method of and motive for committing the crime (level of brutality, existence of a case for self-defense, amount stolen, etc.). This impact on a victim of the intensity of a crime is assumed to decrease over time at an exponential rate.

We assume

$$I_i(t) = c_{j_i} \exp\left[-\frac{1}{\beta_i}(t - t_i) \right],$$

where $I_i(t)$ is the level of injustice due to an unsolved crime i at time t; t_i is the time when the ith crime was committed.

The total injustice due to an unsolved crime is then

$$\int_{t_i}^{\infty} I_i(t)dt = C_{j_i}\beta_i.$$

There are many other possible expressions; we give this as an illustration.

7.3.   Academic Activities

As we change the type of institution being considered, we usually find little change in the type of problems encountered. Our first example relates to policy planning in a university.

Example. University Campus Planning Model [Oliver, Hopkins, and Armstrong]

The purpose of this model is to represent mathematically the relationship among various components of an academic institution. This will be helpful in answering a number of specific policy questions such as: What input flows of new students are required to meet specific output flows of degree winners? How sensitive are enrollment levels to teacher–student ratios?, etc. The formulation of the problem is analogous to an input–output production model for the institution. This may be represented by the linear transformation $Y = AX$, where X is a vector whose components correspond to student degrees of various types and levels; Y is a vector whose components include capital equipment, instructional staff, and also the input stream of new students; A is a matrix whose coefficients include technological requirements such as teacher–student ratio and the fraction of instructional staff derived from the students themselves.

The elements of X and Y are expressed in terms of inventories (e.g., student enrollments and faculty levels), rather than flows.

Let h be the vector of student arrivals into the educational system, whose components represent the students in different educational categories (e.g., lower division, upper division, master's, doctoral, etc.); v be the matrix (v_{ij}), $0 \leqslant v_{ij} \leqslant 1$, where v_{ij} is the fraction of students who enter in the ith category and who subsequently reach the jth category; g be the vector of internal demand for educated students (number per unit time) whose components are students from different categories working as teaching assistants; f be the vector of external demand for educated students (number per unit time) whose components are students leaving the system with different educational levels; L be the vector of student inventories or enrollment levels of students (number); N be the vector of teaching staff inventories; A be the technological requirement matrix (dimensionless); and W be a diagonal matrix of average times required per student to obtain education.

We wish to express the instructional staffing levels in terms of the final demand for trained students.

The average enrollment level is given by

$$L = Vh; \tag{7.1}$$

the average instructional level N is given by

$$N = Wg; \tag{7.2}$$

we also have

$$h = f + g \tag{7.3}$$

and

$$N = AL. \tag{7.4}$$

Substituting (7.1) and (7.2) in (7.3) yields

$$f = V^{-1}L - W^{-1}AL. \tag{7.5}$$

Multiplying both sides of (7.5) by V gives

$$L = [1 - VW^{-1}A]^{-1}Vf. \tag{7.6}$$

Finally, we combine (7.4) and (7.6) to give the required relation

$$N = A[1 - VW^{-1}A]^{-1}Vf.$$

Further, combining (7.1) and (7.6) gives

$$h = V^{-1}[1 - VW^{-1}A]^{-1}Vf.$$

Thus the instructional staffing levels and the required student flows have been obtained.

7.4. Communication and News Transmissions

The same basic situation may be perceived in a variety of ways each of which leads to a different kind of model. Very briefly we give an example of a communication model.

Example. Diffusion Models of News Transmission

Diffusion models describe the diffusion of some object or idea through a population. The unit of adoption may vary from an individual to, for example, a city, or a nation.

A diffusion model takes the form of a differential equation. Both deterministic and stochastic versions have been formulated.

(i) Deterministic:

$$\frac{dn}{dt} = g(n, N),$$

where n is the number of persons who have adopted a given object or idea and N is the total number of persons in the population under consideration.

(ii) Stochastic:

$$\frac{dp_i}{dt} = f(p_i, p_{i-1}, p_{i+1}, N),$$

where p_i is the probability that there are i persons who have adopted the given object or idea at time t, and N is the total number in population.

7.5. Population and Pollution

We now consider models in the area of population and pollution. In our first example we study emigration from the farm to the city.

Example 1. The Relative Population of Cities from Farm Emigration

The general farm population has been emigrating to the cities; we assume that the level of mobility is sufficiently high and that the cost of transportation is negligible in comparison to the cost of the total move so that distance is no object.

There are m cities to which the farm population can go whose population sizes S_i may be ranked as follows: $S_1 > S_2 > \ldots > S_m$. The chance of choosing a city depends on its rank. The first city gets m times as many immigrants as the smallest. Thus, the chance of moving to S_1 are m times as great as those of moving to S_m. An empirical observation is that the tendency of immigrants is to move so that each city gets about the same immigration in relation to its size as any other city. (The rank is used as the weight.) How does this effect the size of cities?

We wish to minimize

$$\sum_{i=1}^{m} i S_i^2$$

subject to

$$\frac{1}{m} \sum_{i=1}^{m} S_i = C,$$

which is a given average population per city. The constraints can be written as

$$\sum_{i=1}^{m} S_i = C_1,$$

where C_1 is the total population of n cities.

We use Lagrange multipliers to obtain $2iS_i + \lambda = 0$.

Thus
$$S_i = \frac{\lambda}{2i}$$

and
$$\lambda = \frac{-2C_1}{\sum_{i=1}^{m} \frac{1}{i}}.$$

Thus

$$S_i = \frac{C_1}{i \sum_{j=1}^{m} \frac{1}{j}} \quad (i = 1, 2, \ldots, m)$$

is the required population of city i.

Our next model refers to the pollution of rivers.

Example 2. River Pollution

In 1963 R. V. Thomann extended a previous model due to D. J. O'Connor by dividing a river estuary along its longitudinal axis into segments flowing into each other and studying the relationship between the amounts of dissolved oxygen (DO) and the biochemical oxygen demand (BOD) from segment to segment. Both of these affect life in the river and also affect the way in which wastes thrown into the river are converted into harmless materials.

Let L_i be the mean concentration of BOD in the ith segment; C_i be the mean concentration of DO in the ith segment; t be the time; V_i be the volume of the ith segment; Q_i be the net waterflow across the upstream boundary of the ith segment; ξ_i be a dimensionless advection factor; E_i be the turbulent exchange factor for the upstream boundary of the ith segment; d_i be the BOD decay rate constant in the ith segment; J_i be the rate of BOD loading to the ith segment from an external source (lb/day); r_i be the reaeration rate of the ith segment (per day); C_{se} be the saturation DO value (lb/ft^3); P_i be any other source or sink of the DO in segment i (lb/day).

They obtained $2n$ equations which described a mass balance of BOD and DO for each of the n segments of the estuary (assuming vertical and lateral homogeneity). These equations may be easily derived and are given by:

$$V_i \frac{dL_i}{dt} = Q_i[\xi_i L_{i-1} + (1-\xi_i)L_i] - Q_{i+1}[\xi_{i+1}L_i + (1-\xi_{i+1})L_{i+1}] + E_i(L_{i-1}-L_i)$$

$$+ E_{i+1}(L_{i+1}-L_i) - d_i V_i L_i + {}^iJ_i, \quad (i = 1, 2, \ldots, n), \tag{7.7}$$

$$V_i \frac{dC_i}{dt} = Q_i[\xi_i C_{i-1} + (1-\xi_i)C_i] - Q_{i+1}[\xi_{i+1}C_i + (1-\xi_{i+1})C_{i+1}] + E_i(C_{i-1}-C_i)$$

$$+ E_{i+1}(C_{i+1}-C_i) - d_i V_i L_i + r_i V_i[C_{sc}-C_i] + P_i \quad (i = 1, 2, \ldots, n) \tag{7.8}$$

7.6. Economic Models

Mathematical models in economics are usually highly sophisticated. Unfortunately, they are often much less accurate in predicting what will occur than in describing what is occurring. Perhaps one reason for this is the fact that economics is a behavioral science and that in economics behavior dominates logic, rather than the reverse, as in queueing theory, for example. Nevertheless, economic models are vitally important and we present several of them.

Example 1. Demand Analysis [Baumol]

A customer has a fixed budget which is to be spent on a number of commodities in such a way as to maximize his total utility. A model for this may be derived as follows:
Let the utility function be

$$U = f(Q_1, Q_2, \ldots, Q_n)$$

and the budget constraint be

$$P_1 Q_1 + P_2 Q_2 + \ldots + P_n Q_n = M,$$

where Q_i is the quantity of commodity i purchased, P_i is the price of commodity i, and M is the total budget available.

Then, using a Lagrange multiplier,

$$U_\lambda = f(Q_1, Q_2, \ldots, Q_n) + \lambda(P_1 Q_1 + P_2 Q_2 + \ldots + P_n Q_n - M),$$

and using $\partial f/\partial Q_i$ as the marginal utility of commodity i, we have

$$\frac{\partial f/\partial Q_i}{\partial f/\partial Q_j} = \frac{P_i}{P_j} \quad \begin{array}{l}(i = 1 \ldots n) \\ (j = 1 \ldots n),\end{array}$$

i.e., the ratio of the marginal utilities of two commodities is the same as the ratio of their prices.

Since $\partial f/\partial Q_i$, $\partial f/\partial Q_j$ will be functions of $Q_1 \ldots Q_m$, solutions of these equations will in general give quantities $Q_1 \ldots Q_n$ for maximum utility.

Example 2. Variation of Income [Shapiro]

National income consists of the sum of three components: (i) consumer expenditures, (ii) induced private investments, and (iii) government expenditures. Data is available at discrete time intervals denoted by the subscript t. Let

$$Y_t = C_t + I_t + G_t,$$

where Y_t is the national income, C_t is the consumer expenditures, I_t is the private investment, and G_t is the government expenditures.

The following assumptions are made relating the variables:

(1) consumer expenditures are proportional to the national income of the preceding period;
(2) private investment in a period is proportional to the increase in consumer expenditures for that period over the preceding period;
(3) government expenditure is the same in all periods.

The problem is now to analyze the behavior of national income under these conditions.

From (1), $C_t = \alpha Y_{t-1}$.
From (2), $I_t = \beta(C_t - C_{t-1})$.
From (3), since G_t is constant, we may choose units such that $G_t = 1$.
Combining equations we obtain

$$\begin{aligned} Y_t &= \alpha Y_{t-1} + \beta(C_t - C_{t-1}) + 1 \\ &= \alpha Y_{t-1} + \beta(\alpha Y_{t-1} - \alpha Y_{t-2}) + 1, \\ Y_t &= \alpha(1+\beta)Y_{t-1} - \alpha\beta Y_{t-2} + 1. \end{aligned}$$

With values assigned to α, β and initial values to Y_{t-2}, Y_{t-1}, an oscillating curve of national income is produced. This curve is either gradually damped to a limit or undergoes increasing oscillations without limit.

Example 3. The Harrod-Domar Growth Theory [Shapiro]

What growth rate in investment input is necessary to produce a given growth rate in output?
Let K be the capital stock, Y be the level of output, Y_p be the potential level of output, and Y_R be the actual level of output.

Define the average capital–output ratio as K/Y. The marginal capital output ratio $\Delta K/\Delta Y$ tells how much additional capital is necessary to provide a specified addition to the flow to output.

In addition, Y/K is the average productivity of capital and $\Delta Y_P/\Delta K$ gives the ratio of increase in potential capacity output to the ΔK increase in capital stock.

It is assumed that

$$\frac{\Delta Y_P}{\Delta K} = \frac{Y}{K} = \text{constant } \sigma.$$

Since in any time period $\Delta K = I$ (investment),

$$\frac{\Delta Y_P}{I} = \sigma \quad \text{or} \quad \Delta Y_P = \sigma I.$$

It is also assumed that

$$\Delta Y_R = \frac{1}{\alpha} \Delta I,$$

where α is the marginal propensity to save.

Now at an equilibrium rate of growth

$$\Delta Y_R = \Delta Y_P = \Delta Y,$$

$$\frac{\Delta I}{\alpha} = \sigma I,$$

and hence

$$\frac{\Delta I}{I} = \alpha \sigma.$$

Further, $\Delta Y = \sigma I = \sigma(\alpha Y)$ in equilibrium, and so

$$\frac{\Delta Y}{Y} = \alpha \sigma,$$

and, therefore,

$$\frac{\Delta I}{I} = \frac{\Delta Y}{Y} = \alpha \sigma.$$

Thus the growth rate of investment and the growth rate of actual output are the same for equilibrium. In addition, the higher the propensity to save, α, the greater the required growth rate, and conversely. The higher the productivity of capital, σ, the greater the required growth rate and conversely.

Our next example is a very practical illustration of economic utilization of a scarce commodity.

Example 4. Economic Utilization of Desalinated Water for an Irrigated Farm – A Linear Programming Model With Uncertain Cost Coefficients [Oak Ridge]

The problem is to determine the size of the farm, the crop pattern, and the schedule for storage and retrieval of water in order to make best use of the output of water in order to make best use of the output of a desalination plant.

Water requirements for the crop start at the planting date, rise with the increased evapotranspiration as foliage is developed, and begin to decline a month or so before the harvest. With high-efficiency sprinkler irrigation in a desert climate, cotton, sunflower, peanuts, soybeans, sorghum, and beans reach a peak water requirement of about 10 in. per month; wheat, potatoes, tomatoes, and citrus reach requirements of about 5 in. per month. Except for beans (three-month) and citrus (year-round), each of these crops has a growing season of about six months. Let $j(1, 2, \ldots, n)$ designate a particular crop schedule suitable for one parcel of land; an example would be beans from June 1st to August 30th and then winter wheat from November 15th to the following May 15th. For parcel j, designate the water requirements in inches for each month $i(i, 2, \ldots, 12)$ as a_{ij}. The annual requirement for parcel j is $A_j = \sum_i a_{ij}$. (For the bean/wheat example, the January to December sequence, rounded to the nearest inch, might be: 2, 3, 5, 6, 1, 4, 10, 6, 0, 0, 1, 2 with a total of $A = 40$.) If X_j is the number of acres in parcel j (planted according to schedule j), then the water volume required in month i by the parcel is $a_{ij}X_j$ acre-inches.

Many coastal locations have reservoirs suitable for underground storage of water. Let s_i designate the volume (acre-inches) stored in month i, and let r_i designate the volume retrieved. The total annual storage is $S = \sum_i s_i$, the total annual retrieval is

$$R = \sum_i r_i, \text{ and } R \le (1-f)S, \tag{7.9}$$

where f is the fraction of the water volume which is lost in storage. Let the desalination plant have a capability of delivering d acre-inches/month and D acre-inches/year. (Because of lost time for maintenance of the desalination plant, $D < 12d$.) The production constraint may be given as

$$d \ge a_{i1}X_1 + \ldots + a_{in}X_n + s_i - r_i \quad (i = 1, 2, \ldots, 12), \tag{7.10}$$
$$D \ge A_1 X_1 + \ldots + A_n X_n + S - R. \tag{7.11}$$

Equations (7.9), (7.10), and (7.11) form a set of 14 constraints. Over the life of this project, income and expenses are not uniform from year to year. For example, certain expenses such as land reclamation do not occur each year, and for citrus crops there is no yield for the first five years of growth. In order to compare income versus expenditures, transactions are financially discounted to a base date. In other words, income and expenses are expressed in terms of their "present worth." Let the present worth of the income associated with parcel j be designated as b_j and the present worth of the expense as c_j, both in $/acre. Income is the product of the yield of the crops on the parcel and their sales price. Expense is the cost for land development, construction of the irrigation system and crop storage facilities, materials such as seed, fertilizer, pesticides, and tractor fuel, and labor for operation and maintenance. In addition, there is a pumping expense of p $/acre-inch associated with retrieval of water from storage. To maximize net income, the objective is thus:

maximize
$$\sum_j (b_j - c_j)X_j - pR. \tag{7.12}$$

The cost of water, i.e., the cost of operating the desalination plant, is deliberately omitted. The cost of operating a dual-purpose plant built for producing both power and desalted water is assumed to remain constant whether or not water is produced. (If water is not produced, more waste heat is generated.) Under such a condition, there is no economic reason for the evaporator not to operate at full capacity. The problem is to make most advantageous use of the water when the plant output is considered fixed. (The cost of operating the evaporator enters the decision of whether or not to build the power plant/evaporator/farm complex but would not affect the design of the farm.)

The market, i.e., the sales price, for some crops is subject to considerable year-to-year fluctuation. It would be prudent, therefore, to solve the problem several times using different sets of sales prices, to indicate the expectation (probability) that each set of prices will occur, and then to decide which farm design is optimal.

Crop yield will respond to change in the quantity of water applied. On the basis of such data, alternative water schedules for particular crops can be tested.

In the model, crops grown sequentially on the same parcel of land were considered independent of crops grown simultaneously on adjacent parcels. Planting, cultivating, and harvesting activities, however, are staggered throughout the year, and joint use can be made of both labor and equipment. Recognition of such joint production economics would require revision of the objective function given by expression (7.12).

It should be noted that instead of maximization of net income, alternative objectives might be selected such as maximization of protein or caloric production.

7.7.   Conflict Resolution

We have been particularly interested in models relating to conflict resolution and in the identification of problems in that area which are amenable to present modeling techniques. We have also been interested in developing new techniques for such problems; we discuss this further in the next chapter.

A fundamental question is that of measurement: How can we determine those aspects of human interaction which are quantifiable in a way which permits the construction of a model?

Measurement may enter conflict resolution in several clearly defined ways. One involves the use of numbers as probabilities; and this involves their use in assigning monetary values to commodities involved. In the latter case, a conflict problem may be analyzed in an economic context of benefits and losses.

Generally, the probability approach to conflict leads to solution in terms of moments (e.g., averages and variances) and in terms of asymptotic stability or the existence of steady-state solutions.

The economic approach also invokes concepts of equilibria and stability. Stability, as used here, connotes a balance of forces which could lead to a military standoff because neither side has more weapons than the other.

Both probabilistic and deterministic models are often based on data which allow determination of their parameters. Such models can be used for prediction and extrapolation.

Game theory is another method which may be used to analyze conflict and which requires the use of measurement. In game theory, the problem is to find a scheme of representing preferences which allows one to conduct an analysis consistent with definitions of stability.

There are two important areas of conflict and conflict resolution which are amenable to modeling. The first involves the analysis of armament and arms races, escalation of hostilities, and the conduct of wars. The second concerns negotiation and bargaining in the settlement of quarrels and hostilities. The main modeling approach used to analyze negotiations is game theory. However, conflicts and their escalation have been studied with a variety of models ranging from algebra and differential equations to control theory as well as game theory, particularly zero-sum games such as the Blotto Game and Submarine Duels. These models may be placed in two categories, descriptive models which analyze stability and normative models which provide means for finding an optimal policy. Most game theoretic analyses attempt to be normative but fall short because of the difficulty of measurement.

It should be evident that mathematical models in conflict resolution serve as an aid to better understanding of the problem rather than a means to obtain the answer to be implemented. The value of general models is actually to improve understanding rather than to give clear-cut numerical answers.

Example 1. Measurement of Power: The Shapley Value [Luce and Raiffa]

Game theory has been analyzed in many different ways. A characteristic function v assigns a value $v(C)$ to every coalition C. (A coalition is a subset of the set N of all players.) A characteristic function is assumed to satisfy the following conditions:

(i) $v(\Phi) = 0$, where Φ is the empty set; and
(ii) $v(C_1 \cup C_2) \geq v(C_1) + v(C_2)$ for $C_1 \cap C_2 = \Phi$ (i.e., superadditivity).

The Shapley value is an *a priori* assessment of the chances of a player in a characteristic function game. It is based on a system of axioms: (i) the efficiency axiom, (ii) the symmetry axiom, (iii) the dummy axiom, and (iv) the additivity axiom. The Shapley value can be described as an average over the marginal contribution of player i, $v(C) - v\{C - (i)\}$ to a coalition C with i in C.

We let $|C|$ be the number of players in C and n be the total number of players. If we take a fixed coalition C, there must be $(|C| - 1)!(n - |C|)!$ sequences of players in order in which player i contributes to C.

Now let L_i denote the set of all C with $i \in C$. The Shapley value for player i is defined as

$$S_i = \sum_{C \in L_i} \frac{(|C| - 1)!(n - |C|)!}{n!} [V(C) - V\{C - (i)\}].$$

This may be regarded as a measurement of the power of player i.

Application to Bargaining Situation [communicated by R. Selten]

Consider the following game of negotiations:

(1) There are five players called small (S).
(2) There are two players called big (B).
(3) The following coalitions formed among these players win the game:

   (a) Two B players and at least one S player.
   (b) One B player and at least three S players.
   (c) Five S players.

The game follows the following cycle:

Step 0. The seven players negotiate among themselves until a winning coalition can be formed with each member of the coalition in agreement with respect to his share of the prize.

Step 1. The coalition is Registered. Ten-minute timer is started.

Step 2. Those players who are not members of the registered coalition try to break up the coalition by offering more attractive coalitions to the members of the registered coalition. All the players can negotiate during this step to protect their positions or to try to improve them. There are two possible results of this step:

(a) The registered coalition stands for 10 minutes. In this case the game is over and the prize is divided as agreed.

(b) The coalition is broken and a new one is formed. The game is then continued from step 1.

The value of this game to each of the players may be computed in a number of ways. One way is to use the Shapley value.

For characteristic functions with a small number of players a simple tabular method can be used in order to compute the Shapley value.

Computation of the Shapley-value for the foregoing seven-person game

This game has the following characteristic function:

$v(i) = 0$ for $i = 1 \ldots 7$;

$v(C) = 27$ if C contains players 1, 2 and at least one other player;

if C contains player 1 or player 2 and at least three of the players 3, 4, 5, 6, 7;

if C contains the players 3, 4, 5, 6, 7;

$v(C) = 0$ in all other cases.

TABLE 7.1. *Computational table for the seven-person demonstration game:*

Coalition type	Number of coalitions with Player 1	$\dfrac{(\lvert C\rvert - 1)!\,(n - \lvert C\rvert)!}{n!}$	Contribution $v(C) - v(C - (i))$	Product of (1), (2), (3)
2 big 1 small	5	$\dfrac{1}{105}$	27	$\dfrac{135}{105}$
2 big 2 small	10	$\dfrac{1}{40}$	27	$\dfrac{270}{140}$
2 big 3 small			0	0
2 big 4 small			0	0
2 big 5 small			0	0
1 big 3 small	10	$\dfrac{1}{140}$	27	$\dfrac{270}{140}$
1 big 4 small	5	$\dfrac{1}{105}$	27	$\dfrac{135}{105}$
1 big 5 small			0	0
5 small			0	0
$s_3 = s_4 = s_5 = s_6 = s_7 = 2\frac{29}{35} \approx 2.83$				$s_1 = 6\frac{3}{7} \approx 6.4$

Some conflicts are resolved by fighting; such situations may be regarded as zero-sum games. In that case each party would attempt to choose its strategy in the confrontation in such a way as to obtain the best return. A simple example follows.

Example 2. Colonel Blotto

Colonel Blotto has four regiments with which he hopes to occupy two posts, but his enemy opposes him with three regiments. If either side has more regiments assigned to a post than the other, he wins by an amount equal to the number of regiments the enemy assigns plus one because a post is considered worth a regiment. The total amount gained or lost is the sum of the payoffs from both posts. If the forces are equal at a post, the payoff is zero at that post. What is the optimal strategy for allocating regiments for Colonel Blotto and for his enemy?

Colonel Blotto has five strategies and the enemy has four. Each strategy is described by an ordered pair of numbers indicating the number of regiments assigned to the first and to the second post respectively. The payoff matrix to Colonel Blotto is as follows:

| | Enemy strategies | | | |
	(3, 0)	(0, 3)	(2, 1)	(1, 2)
(4, 0)	4	0	2	1
(0, 4)	0	4	1	2
Blotto strategies **(3, 1)**	1	−1	3	0
(1, 3)	−1	1	0	3
(2, 2)	−2	−2	2	2

This game has an optimal mixed strategy solution as follows: Blotto should play his respective strategies with probabilities (4/9, 4/9, 0, 0, 1/9); the probabilities for the enemy are (1/18, 1/18, 4/9, 4/9). The value of the game to Blotto is 14/9: if he plays his optimal strategy, he wins at least this number of regiments no matter what the enemy does. We now discuss arms races.

Example 3. Arms Race Model [Richardson]

A very simple but effective model for arms races was given by Lewis Richardson (1881–1953) who developed a model to describe armament buildups between two countries. He assumed (i) that in an armament race between two countries, each country would attempt to increase its armament of the other, (ii) that economic factors impose constraints on armament that tend to diminish the rate of armament by an amount proportional to the size of the existing forces, and (iii) that a nation would build arms, motivated by ambition, grievances, and hostilities even if another nation posed no threat to it.

If the costs of the armament levels of the two sides are $N_1(t)$ and $N_2(t)$, respectively, where t represents time, then the foregoing three conditions may be expressed for each side as:

$$\frac{dN_1}{dt} = kN_2 - aN_1 + g,$$

$$\frac{dN_2}{dt} = lN_1 - bN_2 + h,$$

where k, a, g, l, b, and h are positive constants.

The constants in this model are sometimes referred to in the following way: k, l are called defense or reaction coefficients; a, b are called fatigue or expense coefficients; g, h are called grievance coefficients when positive, good-will coefficients when negative. They are usually assumed to be positive, as suggested in condition (iii) above.

In this model a balance of power is attained when a stable equilibrium at a constant level of expenditure is reached. Stability is obtained when $kl < ab$, that is, the product of the coefficients of reaction to the other side must be less than the product of the coefficients corresponding to the expense of armament. An unstable equilibrium occurs when $ab < kl$ and indicates a runaway arms race.

Now let A and B denote the respective expenditure of two blocks for arms preparation and let A_0 and B_0 be the corresponding cooperative expenditures between the blocks. Put $N_1 = A - A_0$, $N_2 = B - B_0$, and assume for simplicity that $k = l$ and $a = b$. If we substitute these quantities in Richardson's equations and add we obtain

$$\frac{d(A+B)}{dt} = (k-a)\left\{A + B - \left(A_0 + B_0 - \frac{g-h}{k-a}\right)\right\},$$

which says that the rate of change of the combined armament expenditures of both blocks is proportional to the level of the combined expenditures. That is, we have a linear relation.

The last equation describes the arms race of 1909–1913 surprisingly well; Austria–Hungary and Germany were on one side, France and Russia on the other side. Richardson collected data for this period and made a plot as shown in Fig. 7.1.

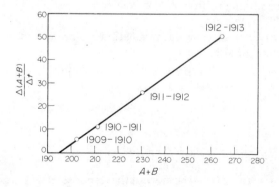

FIG. 7.1.

He drew the straight line suggested by the combined equation and obtained the indicated good fit of the model to the data giving the trend along the line with time. The graph indicates a positive proportionality factor or slope $k - a$. Thus $k - a > 0$ or $k > a$, indicating instability and a runaway race.

In general, it is not easy to obtain an accurate statistical measure of warlike preparations, but such measures are crucial for the credibility of the model. There are other studies in which data have been collected for the purpose of analyzing the stability of the arms race between the United States and the USSR in recent years.

7.8.   Learning

A theory of learning is obviously central to the theory of behavior, for individual behavior is itself largely learned. Although there are several old and well-established theories of learning, the study of this subject has gathered momentum with the formulation of many new models. Much of the credit for this is due to Estes and to Bush and Mosteller, on whose work some of the following examples are based.

Example 1. The Linear Model of Behavior

An individual is exposed to a certain stimulus which evokes one of a number of possible responses and as a result of this response, some event takes place. Over a number of trials, the stimuli become conditioned to certain responses; i.e., they tend to cause those responses by the individual.

Assume that an individual is constrained to two response classes with probabilities p and $1 - p$, respectively. Then a linear recursive equation for p is given by

$$p_{n+1} = \alpha_i p_n + (1 - \alpha_i)\lambda_i, \tag{7.13}$$

where n = number of trials, P_n is the probability that the response of the first class is made on trial n, and α_i, λ_i are two parameters that depend upon the response actually made on trial n and upon the outcome of that trial. $(1 - \alpha_i)$ may be thought of as a rate parameter and λ_i is the asymptote which would be approached if (7.13) were appropriate for every trial beyond a given point in the process. For each event i that could occur on trial n, there corresponds a similar equation.

Bush and Mosteller show that this linear equation can be derived from Estes' theory of stimulus sampling and conditioning.

For more than two response classes, assume a probability vector $\mathbf{p}_n$ for trial n and postulate

$$\mathbf{p}_{n+1} = \tau_n \mathbf{p}_n$$

where τ_i is a linear operator (a transition matrix).

In more general terms, let S be a stochastic operator that describes the effect of the event which has just occurred, i.e., $\mathbf{p}_{n+1} = S\,\mathbf{p}_n$.

Now suppose we wish to treat two or more of the response classes in an identical manner and so we combine them. We can represent this by applying a projection operator C to the probability vectors to obtain $C\mathbf{p}$ and $CS\mathbf{p}$. The stochastic operator now becomes CS, i.e., $(CS)\,C\mathbf{p} = C(S\mathbf{p})$.

Consider again the two-response case

$$p_{n+1} = \alpha_1 p_n + (1 - \alpha_1)\lambda_1 \quad \text{with probability } p_n,$$
$$p_{n+1} = \alpha_2 p_n + (1 - \alpha_2)\lambda_2 \quad \text{with probability } 1 - p_n.$$

A stable asymptotic distribution exists independently of p_0 except when $\lambda_1 = 1$ and $\lambda_2 = 0$. When this is the case, all probabilities are 1 or 0 in the limit. If $f(p_0, \alpha_1, \alpha_2)$ is the probability that a "particle" beginning at p_0 is absorbed in the limit at 1, then

$$f(p_0, \alpha_1, \alpha_2) = p_0 f(\alpha_1 p_0 + 1 - \alpha_1, \alpha_1, \alpha_2) + (1 - p_0) f(\alpha_2 p_0, \alpha_1, \alpha_2).$$

This functional equation has no closed analytic solution.

Various procedures for estimating the parameters of the model from experimental data have been used; it does not appear that any of the parameters are universal constants. The parameter values for a given experiment have not been determined independently of the data from that experiment. $1 - \alpha_i$ is usually observed in the range $0 \leq 1 - \alpha_i \leq 0.2$.

This model has been applied to a rote-learning experiment; i.e., a person memorizes a list of symbols. On each trial the subject is shown the list and then given a recall test. If recall occurs on trial n, the recall probability p_{n+1} on the next trial is given by $p_{n+1} = \alpha p_n + (1 - \alpha)$. This model was found to fit the data well.

We now consider models of behavior viewed as an adaptive process and, in particular, look at behavior in terms of learning and motivation.

Example 2. An Adaptive Behavior Model

Assume an individual performs a learning activity for some specified time over given periods (for example, x could be in hours/day). Let D, the level of difficulty, decrease exponentially with practice. Then $dD/dt = aDx$.

Further, assume that, at any given level of difficulty, practice is pleasurable up to a certain point and then becomes unpleasant.

Let $x = \bar{x}(D)$ be this point. Then motivation may be represented by the equation $dx/dt = -b(x - \bar{x})$. Given initial values, D_0, x_0, at time t_0, the time paths of D and x may be predicted.

This model has not been applied to empirical data, but it does have a number of the required properties and seems theoretically sound.

We now consider how to model an individual's change in opinions over time.

Example 3. Attitude Change Model [Anderson]

At time t an individual may hold one of m opinions. The probability of holding the jth opinion at time $t + 1$ depends upon his opinion i at time t, denoted by $P_{ij}(t + 1)$, where

$$\sum_{j=1}^{m} P_{ij}(t + 1) = 1 \quad (i = 1 \ldots m).$$

The matrix of transition probabilities P is given by

$$p = \begin{bmatrix} P_{11} & \cdots & P_{1m} \\ \vdots & & \\ P_{m1} & \cdots & P_{mm} \end{bmatrix}$$

Therefore, given the individual's initial opinion and the matrix P, it is possible to calculate the probability of any sequence of opinions; it is also possible to find the limiting probabilities for each opinion.

The model was applied to a questionnaire study on voting intentions. *P* was estimated from data collected over a number of months and predictions were made for future opinion changes. The model did not predict well; Anderson attributes this to differences between vocal intent and actual behavior on the part of the voters.

There have been many other learning models; their greatest deficiency has been that they were empirically fitted to data.

Chapter 7—Problems

1. Derive a connection between group size and group performance in solving a problem. In what ways does group size aid in the solution? In what ways does it hinder the finding of a solution? How would you test your model?

2. How is information transmitted among the members of a group? What are the most effective means of transmitting information? Why? What types of models may be used to represent this communication? Can you ensure that all members of the group receive the information? In what time period?

3. How would you decide on the size of prison needed for a given state? [The university planning model will be helpful, but recall that prisoners stay for varying lengths of time.]

4. What forms of pollution exist in your town? How can they be handled? What affects whether or not they will be handled properly? Can you estimate the costs of doing so?

5. How can the courts in your area cut down the time it takes to bring someone to trial? What are the limiting factors? Formulate a simple model for the length of time it takes to bring a case to trial.

6. You have been asked to advise a resort on its tourist industry. How would you proceed? Remember to include not only hotel rooms and facilities, but also access roads, etc. Develop a model to show expected return for given levels of expenditure.

7. Suggest some measures for the grievance and good-will coefficients in Richardson's model of an arms race. How would you test these measures? Repeat for the defense and fatigue coefficients.

8. Formulate a model for group (or gang) formation. What assumptions do you need to make? What would be an optimal size for a group? Why? At what stage will groups tend to break into smaller groups? How would you test your model?

9. You have $20 to spend on dinner. Use the model on demand analysis to decide how you would allocate the money to maximize your satisfaction. Does the model work well? If not, why not?

10. You are teaching your friend how to write a computer program in some given language. How would you test his progress? Would any of the learning models in this chapter be of use? How would you test their validity?

References

Anderson, T. W., Probability models for analyzing time changes in attitudes, in *Mathematical Thinking in the Social Sciences* (ed. P. F. Lazarfeld), Glenco Free Press, 1954.

Baumol, William J., *Economic Theory and Operations Analysis*, Prentice-Hall, Englewood Cliffs, New Jersey, New York, 1961.

Bush, R. R. and C. Mosteller, *Stochastic Models for Learning*, Wiley, New York, 1955.

Defleur, Melvin, *The Flow of Information*, Harper Brothers, New York, 1958, pp. 109–111.

Estes, W. K., A random walk model for choice behavior, in *Mathematical Methods in the Social Sciences* (ed. K. J. Arrow, *et al.*), Stanford University Press, Stanford, 1960.

Oak Ridge National Laboratory Report 4290, *Nuclear Energy Centers; Industrial and Agro-Industrial Complexes*, November, 1968.

O'Connor, Donald J., Oxygen balance of an estuary, *J. Sanitary Enging Div.*, American Society of Civil Engineers, Vol. 86, No. SA3, May 1960.

Oliver, R. M., D. S. P. Hopkins and R. Armstrong, *An Academic Productivity and Planning Model for a university Campus*, University of California at Berkeley, February, 1970.

Richardson, L. F., *Arms and Insecurity*, Borwood, Pittsburgh, 1960.

Shapiro, Edward, *Macro Economic Analysis*, Harcourt, Brace & World, New York, 1966.

Thomann, Robert V., Mathematical model for dissolved oxygen, *J. Sanitary Enging Div.*, American Society of Civil Engineers, Vol. 89, No. SA5, October, 1963.

Bibliography

Bales, R. F., *Interaction Process Analysis*, Addison-Wesley, Reading, Massachusetts, 1951.

Bartholomew, D. J., *Stochastic Models for Social Processes*, Wiley, New York, 1967.

Bartos, O. J., A model of negotiation and some experimental evidence, in *Mathematical Explorations in Behavorial Science*, Illinois, 1965.

Bohigian, Hay E., *The Foundations and Mathematical Models of Operations Research with Extensions to the Criminal Justice System*, Gazette Printers, 1972.

Chaplin, J. P., and T. S. Krawiec, *Systems and Theories of Psychology*, Holt, Rinehart & Winston, New York, 1960.

Coleman, J. S., Reward structures and the allocation of effort, in *Readings in Mathematical Social Science* (ed. P. F. Lazarfeld and N. W. Henry), MIT Press, Cambridge, Massachusetts, 1966.

Horvath, William J., Stochastic models of behavior, in *Management Science*, Vol. 12, August 1966, pp. B513–518.

Horvath, W. J., and C. C. Foster, Stochastic models of war alliances, in *J. Conflict Resolution*, Vol. 7, 1963, pp. 110–116.

Isaacs, R., *Differential Games*, Wiley, New York, 1965.

Kupperman, Robert H., and Harvey A. Smith, Strategies of mutual deterrence, *Shiver*, Vol. 176, April 7, 1972.

Lazarfeld, P. F., and N. W. Henry (eds.), *Readings in Mathematical Social Science*, MIT Press, Cambridge, Massachusetts, 1968.

Luce, R. D. and Raiffa, H., *Games and Decisions*, Wiley, New York, 1957.

Nash, J., Non-cooperative games, *Ann. Math.* Vol. 54, No 2, 1951.

Rashevsky, N., *Mathematical Biology of Social Behavior*, University of Chicago Press, 1959.

Simon, H. A., *Models of Man*, Wiley, New York, 1957.

Simon, H. A., Some strategic considerations in the construction of social science models, in *Mathematical Thinking in the Social Sciences* (ed. P. F. Lazarfeld), Glenco Free Press, 1954.

Sorokin, P., *Social and Cultural Dynamics*, American Bank Co., Vols. I, II, III, 1937, Vol. IV, 1941.

Suppes, P., and R. C. Atkinson, Choice behavior and monetary payoff, in *Mathematical Methods in Small Group Processes* (ed. J. H. Criswell *et al.*), Stanford University Press, Stanford, 1962.

Suppes, P. and R. C. Atkinson, *Markov Learning Models for Multiperson Interactions*, Stanford University Press, Stanford, 1960.

Weiss, Herbert K., Stochastic models of the duration and magnitude to a "deadly quarrel", *Operations Res.*, January 1963.

Chapter 8

Hierarchies and Priorities

8.1. Introduction

In every facet of life we are faced with making a choice among a set of alternatives. Some of these decisions are minor: what clothes to choose each morning, what we wish to eat for breakfast, etc. Other decisions are more important: which job to choose, whom we shall marry, which house we shall buy.

There are, however, decisions which affect many others beside ourselves. The managing director of a large company has to make many important decisions: where to locate a new plant, which products to concentrate on, which people to promote, how to apportion profits, etc. Public officials have to make major resource allocation decisions and other decisions about policies affecting thousands of people.

Even in situations where the decision about a desired final outcome has been made, decisions about intermediate policies to achieve this outcome are necessary.

All of this points up the necessity of a method for making decisions, particularly where a number of objectives have to be satisfied. The method of analytical hierarchies gives us a means of making such decisions in a rational manner.

Essentially there have been three ways of using logic to explain the phenomena of the real world. The first and time-honored method is that of cause and effect. It is known as the reductionist method upon which most scientific research so far has been based. The next is probabilistic and utilizes understanding of random process to explain occurrences by averages and deviations from them. The third and most recent is the systems approach which is concerned with the interactions of the parts within a system and the interactions of the system with its environment. This method of understanding is synthetic and looks at the overall purposes governing the design and functions of a system in order to explain its behavior. The systems approach is hierarchic in its nature, and goes from the particular to the general. Although synthesis cannot be separated from analysis and causality, it is different in its general outlook. Purpose and its fulfillment is its primary concern. Obviously then, priorities in the fulfillment of purpose become essential. Thus, the representation of a system according to its purposes, its environmental constraints, its actors, their objectives, the functions of the system, and the parts which perform these functions takes on a hierarchical form. One would then be concerned with the priority impact of any of these elements on the overriding purpose of the system.

8.2. Paired Comparisons

We first assume that n activities are being considered by a group of interested people and that the group's goals are:

(a) to provide judgments on the relative importance of these activities;

(b) to ensure that the judgments are quantified to an extent which also permits a quantitative interpretation of the judgments among all activities.

Clearly, goal (b) will require appropriate technical assistance.

Our goal is to describe a method for deriving, from the group's quantified judgments (i.e., from the relative values associated with *pairs* of activities), a set of weights to be associated with *individual* activities; in a sense defined below, these weights should reflect the group's quantified judgments. What this approach achieves is to put the information resulting from (a) and (b) into usable form without deleting information residing in the qualitative judgments.

Let $C_1, C_2, \ldots, C_n$ be the set of activities. The quantified judgments on pairs of activities C_i, C_j are represented by an n-by-n matrix

$$A = (a_{ij}), \quad (i, j = 1, 2, \ldots n).$$

Suppose now we have n elements and we wish to compare them. We create the matrix of relative weights by asking the question: Of two elements i and j, which is more important with respect to some given property and how much more? We relate the importance of j by using the scale shown in Table 8.1 to give the values of a_{ij}.

TABLE 8.1 *The scale and its description*

Intensity of importance	Definition	Explanation
1*	Equal importance	Two activities contribute equally to the objective
3	Weak importance of one over another	Experience and judgment slightly favor one activity over another
5	Essential or strong importance	Experience and judgment strongly favor one activity over another
7	Demonstrated importance	An activity is strongly favored and its dominance demonstrated in practice
9	Absolute importance	The evidence favoring one activity over another is of the highest possible order of affirmation.
2, 4, 6, 8	Intermediate values between the two adjacent judgments	When compromise is needed
Reciprocals of above numbers	If activity i has one of the above numbers assigned to it when compared with activity j, then j has the reciprocal value when compared with i	
Rationals	Ratios arising from the scale	If consistency were to be forced by obtaining n numerical values to span the matrix

*On occasion in 2-by-2 problems, we have used $1 + \varepsilon, 0 < \varepsilon \leqslant \frac{1}{2}$ to indicate very slight dominance between two nearly equal activities.

This means that we let $a_{ij} = 5$ if element i (or activity i) is essentially more important than activity j. Suppose we carry out this process to create the first row of the matrix A. If our judgments were completely consistent, the remaining rows of the matrix would then be completely determined (and the entries would not, in general, be integers). However, we do not assume consistency other than by setting $a_{ji} = 1/a_{ij}$. We repeat the process for each row of the matrix, making independent judgments over each pair.

It turns out in practice that this scale gives a workable representation of the way in which people think and compare similar elements [Saaty, *J. Math. Psych*, 1977].

Our concern now is to decide how the elements at each level should be weighted. We consider first a simpler problem. Assume that we are given n stones $A_1 \ldots A_n$, whose weights $w_1 \ldots w_n$, respectively, are known to us. Let us form the matrix of pairwise ratios whose rows give the ratios of the weights of each stone with respect to all others. Thus we have the matrix:

$$
A \equiv
\begin{array}{c|cccc}
 & A_1 & A_2 & \cdots & A_n \\
\hline
A_1 & \dfrac{w_1}{w_1} & \dfrac{w_1}{w_2} & \cdots & \dfrac{w_1}{w_n} \\[2ex]
A_2 & \dfrac{w_2}{w_1} & \dfrac{w_2}{w_2} & \cdots & \dfrac{w_2}{w_n} \\[1ex]
\vdots & \vdots & \vdots & & \vdots \\[1ex]
A_n & \dfrac{w_n}{w_1} & \dfrac{w_n}{w_2} & \cdots & \dfrac{w_n}{w_n}
\end{array}
\begin{pmatrix} w_1 \\ w_2 \\ \vdots \\ w_n \end{pmatrix}
= n \begin{pmatrix} w_1 \\ w_2 \\ \vdots \\ w_n \end{pmatrix}
$$

We have multiplied A on the right by the vector of weights w. The result of this multiplication is nw. Thus, to recover the scale from the matrix of ratios, we must solve the problem $Aw = nw$ or $(A - nI)w = 0$. This is a system of homogeneous linear equations. It has a nontrivial solution if and only if the determinant of $(A - nI)$ vanishes, i.e., n is an eigenvalue of A. Now A has unit rank since every row is a constant multiple of the first row and thus all its eigenvalues except one are zero. The sum of the eigenvalues of a matrix is equal to its trace and in this case, the trace of A is equal to n. Thus n is an eigenvalue of A and we have a nontrivial solution. The solution consists of positive entries and is unique to within a multiplicative constant.

To make w unique we normalize its entries by dividing by their sum. Thus given the comparison matrix we can recover the scale. In this case the solution is any column of A normalized. Note that in A we have $a_{ji} = 1/a_{ij}$ the reciprocal property. Thus, also, $a_{ii} = 1$. Also, A is consistent, i.e., its entries satisfy the condition $a_{jk} = a_{ik}/a_{ij}$.

Thus the entire matrix can be constructed from a set of n elements which form a spanning tree across the matrix.

In the general case we cannot give the precise values of w_i/w_j but estimates of them. For the moment let us consider an estimate of these values by an expert who we assume makes small errors in judgment. From matrix theory we know that small perturbation of the coefficients implies small perturbation of the eigenvalues. Our problem now becomes $A'w' = \lambda_{max} w'$ where λ_{max} is the largest eigenvalue of A'. To simplify the notation we shall continue to write $Aw = \lambda_{max} w$ where A is the matrix of pairwise comparisons. The problem now is how good is the estimate of w. Note that if we obtain w by solving this problem the matrix whose entries are w_i/w_j is a consistent matrix which is our consistent estimate of the matrix A. A itself need not be consistent. In fact, the entries of A need not even be ordinally consistent, i.e., A_1 may be preferred to A_2, A_2 to A_3, but A_3 is preferred to A_1. What we would like is a measure of the error due to inconsistency. It turns out that A is consistent if and only if $\lambda_{max} = n$ and that we always have $\lambda_{max} \geq n$. This suggests using

$\lambda_{\max} - n$ as an index of departure from consistency. But

$$\lambda_{\max} - n = - \sum_{i=2}^{n} \lambda_i, \quad \lambda_{\max} = \lambda_1,$$

where λ_i, $i = 1, \ldots, n$, are the eigenvalues of A. We adopt the average value $(\lambda_{\max} - n)/(n-1)$ which is the (negative) average of λ_i, $i = 2, \ldots, n$ (some of which may be complex conjugates).

Now we compare this value for what it would be if our numerical judgments were taken at random from the scale $1/9, 1/8, 1/7, \ldots, 1/2, \ldots, 1, 2, \ldots, 9$ (preserving the reciprocal relationship in order to improve consistency).

We have for different order random matrices and their average consistencies:

Size of matrix	1	2	3	4	5	6	7	8	9	10
Random consistency	0.00	0.00	0.58	0.90	1.12	1.24	1.32	1.41	1.45	1.49

One rarely has to go beyond 7×7 matrices (in order to maintain connectedness between relations and keep the consistency relatively high). It is with these numbers that we divide the consistency index and recommend revisions if the ratio is considerably higher than 10%. This gives the consistency ratio CR.

We note that solution of the largest eigenvalue problem, when normalized, gives us a unique estimate of an underlying ratio scale.

8.3. The Calculation of the Weights and Priorities

We have noted that the normalized principal eigenvector of the paired comparison matrix A gives the weights of the elements being compared. An efficient program for finding this vector has been used in our applications. There is, however, a very simple method for finding an approximation to the vector of weights.

Suppose that the matrix A takes the following form:

$$\begin{array}{c} \\ x \\ y \\ z \end{array} \begin{array}{ccc} x & y & z \\ \begin{bmatrix} 1 & 4 & 9 \\ 1/4 & 1 & 2 \\ 1/9 & 1/2 & 1 \end{bmatrix} \end{array}$$

where the ratios are not maintained exactly. Normalizing the first column by dividing each element by the sum of the elements in the column gives $\begin{pmatrix} 0.735 \\ 0.184 \\ 0.082 \end{pmatrix}$, normalizing the second

column gives $\begin{pmatrix} 0.727 \\ 0.182 \\ 0.091 \end{pmatrix}$, while the third column normalizes to $\begin{pmatrix} 0.750 \\ 0.167 \\ 0.083 \end{pmatrix}$. Each of these columns can be regarded as a measure of the relative weights. A "reasonable" method of evaluating the weights would be to take the three columns

$$
\begin{matrix}
0.735 & 0.727 & 0.750 \\
0.184 & 0.182 & 0.167 \\
0.082 & 0.091 & 0.083
\end{matrix}
$$

and to average across the rows to give

$$
W = \begin{pmatrix} 0.737 \\ 0.178 \\ 0.085 \end{pmatrix}
$$

and to use this "average" column for the relative weights.

We can then multiply Aw to obtain an estimate of $\lambda_{max} w$. We then divide each component of $\lambda_{max} w$ by the corresponding component of w to obtain an estimate of λ_{max}; we may average over these estimates to obtain an overall estimate of λ_{max}. We then compute the consistency index $(CI) = (\lambda_{max} - n)/(n - 1)$.

This can be compared with its value derived from random entries and the vector w is accepted if the ratio CR is of the order of 10% or less.

Another quick way to obtain an estimate of the principal eigenvector is to take the geometric mean of each row and then normalize the elements of the resulting column. With perfect consistency one can also add the elements in each row and normalize the resulting column, but in general this is an unsatisfactory way.

Example. World Influence of Nations [Saaty and Khouja]

A number of people have studied the problem of measuring world influence of nations. We have briefly examined this concept within the framework of our model. We assumed that influence is a function of several factors. We considered five such factors: (1) human resources; (2) wealth; (3) trade; (4) technology; and (5) military power. Culture and ideology, as well as potential natural resources (such as oil), were not included.

Seven countries were selected for this analysis. They are the United States, the USSR, China, France, the United Kingdom, Japan, and West Germany. It was felt that these nations as a group comprised a dominant class of influential nations. It was desired to compare them among themselves as to their overall influence in international relations. We realize that what we have is a very rough estimate—mainly intended to serve as an interesting example of an application of our approach to priorities. We shall illustrate the method with respect to the single factor of wealth.

In Table 8.2 we give a matrix indicating the pairwise comparisons of the seven countries with respect to wealth. For example, the value 4 in the first row indicates that influence through wealth is between weak and strong importance in favor of the United States over the USSR. The reciprocal of 4 appears in the symmetric position, indicating the inverse relation of relative strength of the wealth influence of the USSR compared to the United States.

Note that the comparisons are *not* consistent. This basic inconsistency cannot be eradicated merely by changing scale. Moreover, there are also apparent numerical (or scale) inconsistencies. For example, the United States : China = 9 (not 28) despite the fact that the United States : USSR = 4 and USSR : China = 7.

TABLE 8.2. *Wealth*

	United States	USSR	China	France	United Kingdom	Japan	W. Germany
United States	1	4	9	6	6	5	5
USSR	1/4	1	7	5	5	3	4
China	1/9	1/7	1	1/5	1/5	1/7	1/5
France	1/6	1/5	5	1	1	1/3	1/3
United Kingdom	1/6	1/5	5	1	1	1/3	1/3
Japan	1/5	1/3	7	3	3	1	2
W. Germany	1/5	1/4	5	3	3	1/2	1

Nevertheless, when the requisite computations are performed, we obtain relative weights of 43.3 and 21.7 for the United States and the USSR, and these weights are in striking agreement with the corresponding GNPs as percentages of the total GNP in Table 8.3. Thus, despite the apparent arbitrariness of the scale, the irregularities disappear and the numbers occur in good accord with observed data.

Compare the normalized eigenvector column derived by using the matrix of judgments with the actual GNP fraction given in the last column. The two are very close in their values. Estimates of the actual GNP of China range from 74 to 128 billion.

TABLE 8.3. *Normalized wealth eigenvector*

	Normalized Eigenvector	Actual GNP* (1972)	Fraction of GNP Total
United States	0.429	1167	0.413
USSR	0.231	635	0.225
China	0.021	120	0.043
France	0.053	196	0.069
United Kingdom	0.053	154	0.055
Japan	0.119	294	0.104
W. Germany	0.095	257	0.091
Total		2823	

*Billions of dollars.

8.4. Hierarchies in Decision Making

We shall not attempt to give a full theoretical development of the method of analytical hierarchies here, but will (in a later section) explain briefly why and how hierarchies work, and will then show how to use the method in practice.

The essential requirement for analysis by hierarchies is that one should be able to decompose a problem into levels, where each level consists of similar elements and has an impact on the levels above and below. We have found in practice that this is the way in which one instinctively structures a problem.

Consider a typical problem. Some decision will have to be made: this is placed at the apex of the hierarchy. Those people or groups who have an interest in the problem and in the final outcome will provide the next level of the hierarchy. Each of these groups will have a number of objectives which are of different degrees of importance to them. (Some of these objectives may be held in common.) The objectives will provide the next level. There are also the possible decisions or outcomes to be evaluated. Each of these will satisfy the objectives to a greater or lesser extent and, in consequence, they provide the final level of the hierarchy.

A typical hierarchy is shown in Fig. 8.1.

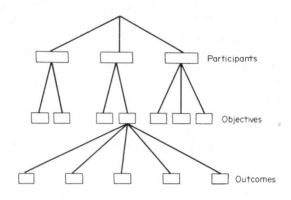

FIG. 8.1.

8.5. Weighting through the Hierarchy

We may now apply the weighting process to the participants at the first full level of the hierarchy. By forming the matrix of paired comparisons of the power of each party to affect the outcome, we are able to derive the normalized eigenvector to obtain the power of each of the parties. The process may be repeated for the objectives of each party. A typical judgment which has to be made here is: For a given party, which of two of its objectives is it likely to pursue more, and how much more? This is a process of evaluating the objectives in pairs according to their contribution to each element, a party, in the level immediately above. A vector of objective weights is obtained for each party.

The final outcomes are now similarly weighted by a comparison matrix according to how well they would satisfy a given objective in the view of the party whose objective is under consideration. The process is repeated for each objective of each party.

In making the judgments about the power of a party to influence a decision, or about the relative importance of its objectives, etc., one either should seek advice from those who are actively involved or who are particularly knowledgeable about the situation, or should form one's own judgments based on a complete background study. The former is obviously preferable, but is not always possible.

The weights for each of the outcomes may now be obtained by composite weighting through the hierarchy. We follow a path from the outcome at the apex to each final outcome at the lowest level and multiply the weights along each segment of the path. The result is a normalized vector of final weights for the possible outcomes under consideration.

There are two possible (and related) ways to view this process. One is to consider a flow of power taking place downward along the paths of the hierarchy from the initial source; this flow splits according to the power of the parties. Each party then divides its available power among its objectives; this accounts for the first multiplication. The flow to each objective is further divided to allow for the way in which the outcomes satisfy each objective; this accounts for the second multiplication. Each outcome receives a contribution from each objective; these are added to give a final weight for each outcome. The initial unit flow has then moved down to the final outcomes to show how much power would be transferred to each. The second approach is to regard the weights as probabilities of independent events, so that the final weights indicate the probabilities that each final outcome will take place if each party used its power to satisfy its objectives.

The outcome to receive the highest weight then indicates an optimum which may be either a most likely or a most desirable result.

This is, of necessity, a very simplified account of the process, although it captures the essentials and indicates how the method works in practice. The interested readers will find a detailed exposition of the theoretical underpinning and of many ramifications and interesting results in the references.

Example. Job Decision

A student who had just received his PhD was interviewed for three jobs (*A*, *B*, and *C*). His criteria (Fig. 8.2) for selecting the jobs and their pairwise comparison matrix are given along with the eigenvector (normalized) associated with the maximum eigenvalue (Table 8.4). In this matrix, the comparisons were made relative to which factor of a pair was more important (and by how much) when considering overall satisfaction with a job. Thus the value of 5 in row three column four indicates that benefits are strongly more important than colleague associations.

TABLE 8.4

Overall satisfaction with job	Research	Growth	Benefits	Colleagues	Location	Reputation	Eigenvector
Research	1	1	1	4	1	1/2	0.18
Growth	1	1	2	4	1	1/2	0.19
Benefits	1	1/2	1	5	3	1/2	0.19
Colleagues	1/4	1/4	1/5	1	1/3	1/3	0.05
Location	1	1	1/3	3	1	1	0.12
Reputation	2	2	2	3	1	1	0.30

Now each job is compared with each of the other jobs with respect to how they bear upon each of the six criteria (Table 8.5).

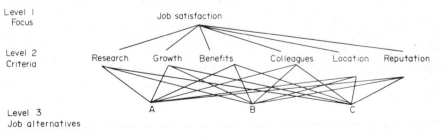

FIG. 8.2.

TABLE 8.5

	Research				Growth				Benefits		
	A	*B*	*C*		*A*	*B*	*C*		*A*	*B*	*C*
A	1	1/4	1/2	*A*	1	1/4	1/5	*A*	1	3	1/3
B	4	1	3	*B*	4	1	1/2	*B*	1/3	1	1
C	2	1/3	1	*C*	5	2	1	*C*	3	1	1

	Colleagues				Location				Reputation		
	A	*B*	*C*		*A*	*B*	*C*		*A*	*B*	*C*
A	1	1/3	5	*A*	1	1	7	*A*	1	7	9
B	3	1	7	*B*	1	1	7	*B*	1/7	1	5
C	1/5	1/7	1	*C*	1/7	1/7	1	*C*	1/9	1/5	1

The eigenvectors (one for each criterion) just obtained are each weighted by the eigenvector component of the associated criterion and the results summed and normalized. We obtain:

$$0.16 \begin{bmatrix} 0.14 \\ 0.63 \\ 0.24 \end{bmatrix} + 0.19 \begin{bmatrix} 0.10 \\ 0.33 \\ 0.57 \end{bmatrix} + - - + 0.30 \begin{bmatrix} 0.77 \\ 0.17 \\ 0.05 \end{bmatrix} \longrightarrow \begin{bmatrix} 0.40 \\ 0.34 \\ 0.26 \end{bmatrix} \begin{matrix} A \\ B \\ C \end{matrix}$$

We have prioritized the jobs. The differences were sufficiently large for the candidate to accept the offer to job *A*.

The previous example was essentially a one-person conflict problem. Examples of conflicts and decisions involving more than one person may be found in the references.

There is another side to these problems. Suppose that a party to a conflict does not like the projected final outcome. Can he do anything to change this outcome? This question is examined next?

8.6. The Forward and Backward Processes in Planning and in Conflict Analysis

It is important to identify those outcomes which are likely to emerge and which, in large measure, satisfy the objectives of each party to a conflict or decision process. Such a process, which is descriptive, may be regarded as a one-point boundary problem fixed at the present state. It is the forward process. Given the present actors and their current objectives, capabilities, and policies, which outcome is the most likely to emerge? This outcome may be a composite of a number of pure outcomes which have been considered.

There is, however, an alternative approach to the solution of a conflict problem. Given a desired future outcome, what can or should be done to achieve such an outcome? Working backward, one assesses the problems and opportunities which affect this outcome and identifies those policies which would be most effective in producing the desired outcome. Such an approach, which is normative, may be regarded as a one-point boundary problem fixed at the future. It is the backward process.

A combination of the forward and backward processes is frequently used in planning. In order to apply the backward process, it is first necessary to find the desired outcome for each of the parties in the problem, and to evaluate their reactions to all of the outcomes. To do this, one may use the weights for the outcomes which were obtained from the first forward process and note how much of each weight came from each party.

The forward process indicates which outcome is likely to emerge, given the existing actors, objectives, and policies. In the backward process one tries to find ways in which a desired outcome may be made acceptable to those who do not give it a high weighting. This may enable a party which strongly desires a given outcome to see what it must do to induce other parties to move toward the desired outcome. The new policies which this process suggests are adjoined to the original policies and the forward process is repeated. The desired outcome then may be modified by some of the features of the newly emergent outcome and the process may be iterated.

If one of the parties realizes that its weighting is not very high and that, in consequence, it cannot influence the outcome as much as it would desire, that party may take action to increase its power by alliances or by the use of threat options, including a credible increase in determination matched by effective action.

This forward and backward process has been applied in a number of examples. [Alexander and Saaty, Emshoff and Saaty, McConney and Vrchota].

8.7. Design of a transport system for the Sudan: Priorities—Investment

Against a background of great potential agricultural riches, the Sudan, the largest country in Africa (967,491 square miles) but with only an 18.2 million population, is today a poor country with a GNP of about 2.8 billion dollars. Oil countries in the Middle East and international agencies, including the World Bank, recognize the capacity of the Sudan as a major provider of food for Africa and the Middle East, and have been investing in its development (see Saaty, 1977).

Incidentally, the oil-rich Arab countries' populations do not exceed 20 million and, hence, their need does not even begin to make a difference in how many people the Sudan can feed. Even if its northern neighbor, Egypt, were to be included for one-half of its population (estimated at 50 million by 1985) to be fed by Sudan, there would still be land to feed perhaps 100 million more people. The entire economy of the Sudan and, in particular, the agricultural sector, suffer from lack of adequate transportation.

The Sudan is serviced by four major modes of transportation: rail, road, river, and air. These modes are combined together to provide a sparse and far-flung transportation infrastructure. The air network is centered at Khartoum and the rail and road systems are oriented for export through Port Sudan. The country is characterized by low transport connectivity. The object was to develop a transport plan for the Sudan by 1985.

The functions of a system can be represented by a hierarchy with the most important "driving" purposes occupying the top level and the actual operations of the system at the lowest level. In the case of the Sudan, overall development occupied the top level, followed by a level of scenarios or feasible outcomes of the future. The third level consisted of the regions of the Sudan as it was desirable to know their impact on the scenarios and in turn the impact of the scenarios on the top level. The fourth level consisted of the projects for which it is desired to establish their priorities by studying their impacts on the regions. Thus, we had a four-level hierarchy in the study. By composing the impacts of the fourth level on the third, the third on the second, and the second on the first, we obtain the overall impact of each project in the fourth level on the overall development of the Sudan represented in the first.

Pairwise comparison of the four scenarios according to their feasibility and desirability by 1985 gave rise to the matrix presented in Table 8.6.

TABLE 8.6. *Priorities of the scenarios*

		I	II	III	IV
Status quo	I	1	1/7	1/5	1/3
Agricultural export	II	7	1	5	5
Balanced regional growth	III	5	1/5	1	5
Arab-African regional expansion	IV	3	1/5	1/5	1

The priorities of the scenarios in the order they are listed are: 0.05, 0.61, 0.25, 0.09. As can be seen, scenario II dominates, with scenario III next in importance. Since the future is likely to be neither one nor the other, but rather a composition of these scenarios—with emphasis indicated by the priorities—this information was used to construct a composite scenario of the Sudan of 1985. This scenario is intended as the anticipated actual state of the future, it being a proportionate mix of the forces which make up the four scenarios described above. The composite scenario takes the main thrust of scenario II, the future given by far the highest priority, and is enlarged and balanced with certain elements from scenarios III and IV. This composition indicates the likelihood of a synergistic amplification of individual features.

The Sudan has 12 regions whose individual economic and geographic identity more or less justifies political division into distinct entities. The regions were compared pairwise in separate matrices according to their impact on each of the scenarios. They comprise the third hierarchy level. The resulting eigenvectors are used as the columns of a matrix which, when multiplied by the eigenvector of weights or priorities of the scenarios, gave a weighted average for the impact of the regions (Table 8.7).

TABLE 8.7. *Priority weights of regions* (%)

Bahr El Ghazal	3.14	Khartoum	21.40
Blue Nile	6.55	Kordofan	5.96
Darfur	5.37	Northern	2.94
East Equatoria	1.70	Red sea	22.54
Gezira	12.41	Upper Nile	3.37
Kassala	5.25	West Equatoria	9.39

Now the projects, of which there were 103 determined according to GNP growth rates which suggest supply–demand and flow of goods, comprised the fourth level of the hierarchy. They were compared pairwise in 12 matrices according to their impact on the regions to which they physically belonged. A project may belong to several regions and this had to be considered. The resulting matrix of eigenvectors was again weighted by the vector of regional weights to obtain a measure of the overall impact of each project on the future. This gave rise to the following kind of Table (8.8) of which there are nine in the final report:

We examined the 4.3%, the present growth rate, and found that most of the current

TABLE 8.8. *The transportation development plan: phase I (1974 price level in £S000,000) (6% GNP growth rate)*

Projects	Distance (km)	Priority	Class GNP rates 4.3% L	4.3% H	6.0% L	6.0% H	7.3% L	7.3% H	Cost A	Cost B	Cost C	Recommended class	Flow	Other	Committed (financing in progress)	Total	Foreign currency	Local currency
RAIL																		
Port Sudan–Haiya	203	4.724	A	B	A	B	A	B	9.10	7.10	–	A	X			9.10	4.55	4.55
Haiya–Atbara	271	3.455	B	B	B	B	A	B	12.20	9.50	–	B	X			9.50	6.30	3.20
Atbara–Khartoum	313	8.443	B	B	B	B	B	B	14.10	11.00	–	B	X			11.00	7.30	3.70
El Rahad–Babanusa	363	1.005	B	B	B	B	B	B	–	12.70	–	B	X			12.70	8.50	4.20
Fleet (6% GNP)																40.90	40.90	–
Maintenance facilities																2.00	1.00	1.00
sub-total																85.20	68.55	16.65
ROAD																		
Wad Medani–Gedaref	231	2.840	A	A	A	A	A	A	23.90	–	–	A	X		X	23.90	16.70	7.20
Gedaref–Kassala	218	0.872	A	A	A	A	A	A	14.20	–	–	A	X		X	14.20	9.90	4.30
Kassala–Haiya–Port Sudan	625	2.229	A	A	A	A	A	A	50.00	–	–	A	X		X	50.00	35.00	15.00
Wad Medani–Sennar	100	0.526	A	A	A	A	A	A	14.90	–	–	A	X		X	14.90	10.40	4.50
Sennar–Kosti	110	0.345	A	A	A	A	A	A	7.20	–	–	A	X		X	7.20	5.00	2.20
Sennar–Es Suki	47	0.546	A	A	A	A	A	C	7.00	–	–	A	X			7.00	4.90	2.10
Ed Dubeibat–Kadugli	137	1.253	C	C	C	C	B	C	–	12.30	8.80	B	X		X	12.30	7.40	4.90
Kadugli–Talodi	100	0.266	–	–	–	B	–	C	–	6.60	–	–						
Nyala–Kass–Zalingei	210	0.951	B	C	B	C	C	A	–	11.30	7.40	B	X			11.30	6.80	4.50
Jebel Al Aulia–Kosti*	300	1.567	B	B	B	B	A	B	44.70	29.70	–	–	X	High cost, alternative provided				
Juba–Himuli	190	0.329	C	C	C	C	C	C	–	8.70	5.30	C	X			5.30	1.60	3.70
Juba–Amadi–Rumbek–Wau	725	0.494	C	C	C	C	C	C	–	–	20.30	C	X	Together with alternate, high priority		20.30	6.10	14.20
Fleet																20.80	20.80	–
sub-total																187.20	124.60	62.60

* The priority rating of this project is based mostly on potential rather than present development. In view of its high cost relative to other road projects, it has been omitted. It is recommended that it be given urgent consideration in the following planning period.

facilities with the prevailing level of efficiency would be crammed to their limit. Obviously, a compromise with a rational justification for growth had to be made somewhere between these two extremes. When we examined the 6% GNP growth rate, found feasible by the econometric analysis, it provided excellent guidelines for those projects which were found to be needed at 4.3% and remained invariant with high priority at 7.3%. These were mostly the projects we recommended for implementation. The ratios of priorities to costs served as a measure of the effectiveness of investment. Six billion dollars have been earmarked for expenditure in the Sudan over the next few years closely following the recommendation of the study.

Total investment requirements to achieve the composite scenario projected growth of real GNP at 4.3%, 6%, and 7.3% per year are given in Table 8.9. For example, at 7.3% they are estimated to be approximately $5105 million at 1974 price levels, or $7647 million at current price levels (considering inflation between 1974 and 1985). The latter figure represents approximately 10% of the GNP each year over the planning period, 1972–1985. This will be divided among the major sectors as shown.

TABLE 8.9. *Dollars (millions), current prices*

	4.3%	6%	7.3%
Transport	978.33	1789.88	2899.96
Agriculture	1372.75	1695.53	2183.30
Industry	588.28	963.33	1456.08
Services	1307.35	1194.58	1107.20
Total	4246.71	5643.32	7646.54

8.8.   Rationing Energy to Industries: Optimization

Only a short time ago it was unthinkable and deemed as an academic exercise to speak of rationing energy. It was felt that there could not be a crippling energy crisis because our energy czars and planners would presumably take our needs into their projections. Today, things look very different. Witness the cases of the lack of natural gas in the cold winter of 1976–1977 which caused the shut down of some schools and industries and the coal strike of 1977–1978 (See Saaty and Mariano).

In any case we must face the needs of our homes, offices, industries, and massive transportation systems not simply by making additional supplies of energy available but also by replanning and redesign to improve efficiency and to diminish the 7% annual rise in energy consumption.

Besides improving efficiency, for the long range we need to consider several alternatives to prevent severe energy shortages. Among them are:

(1) Reduction of US consumption to the level of domestic oil production.
(2) Discovery or development of new forms of energy such as coal gasification, geothermal, nuclear fission, nuclear fusion, and solar. But these forms are now in short supply.
(3) Rationing. Although rationing is not an attractive alternative, we have seen that in

cases of severe weather, energy had to be diverted from schools and industries in the Midwest to accommodate homeowner needs. Rationing can become a pressing alternative if supplies dry up or continuity in importing oil is seriously threatened.

In this application, we confined our analysis to manufacturing industries. Examples of the groups we considered are: (1) food and kindred products, (2) tobacco manufacturers, (3) textile mill products, (4) apparel and related products, (5) lumber and wood products, (6) furniture and fixtures, (7) paper and allied products, (8) printing and publishing, (9) chemicals and allied products, (10) petroleum and coal products, etc.

The optimal weights generated for these classes of industries are applicable on a yearly basis and, therefore, the actual scheduling of allocation on a day-to-day or a month-to-month basis is not made explicit by the model. Our approach can be extended to peak power demand considerations where shortage of power may occur in a short time duration. In this case, the optimal scheduling of power and its allocation will be determined as a function of time.

We used the following objectives which fall into two classes: class 1, characterized by two measurable indicators; contribution to economic growth (measured in dollars), and impact on the environment (measured in tons of pollutants); class 2, characterized by three qualitative indicators; contribution to national security, to health, and to education. The measures for these were derived using judgments and the eigenvalue procedure.

The results of the two classes were composed hierarchically to obtain an overall priority for each industrial group.

As the real-life problem is too long to work out here we have chosen an example which illustrates how one does a rationing problem. It combines priorities and optimization.

The problem in the energy demand allocation is concerned with finding allocation weights for several large users of energy according to their overall contribution to different goals of society. Let us assume the following conditions.

There are three large users of energy in the United States: C_1, C_2, and C_3. The goals against which these energy users will be evaluated are: contribution to economic growth, contribution to environmental quality, and contribution to national security. Based on the overall objective of social and political advantage the matrix of paired comparisons of these three goals on the previously described scale from 1 to 9 is given by:

$$
\begin{array}{cccc}
 & \text{Economic} & \text{Environment} & \text{National security} \\
\text{Economic} & \begin{pmatrix} 1 \\ 1/5 \\ 1/3 \end{pmatrix} & \begin{matrix} 5 \\ 1 \\ 5/3 \end{matrix} & \begin{matrix} 3 \\ 3/5 \\ 1 \end{matrix} \end{pmatrix} \\
\text{Environment} & & & \\
\text{National Security} & & &
\end{array}
$$

The normalized eigenvector corresponding to the dominant eigenvalue $= 3$ of this matrix is given by:

$$
P(0) = \begin{pmatrix} 0.65217 \\ 0.13044 \\ 0.21739 \end{pmatrix}
$$

The decision-maker, after a thorough study, has made the following assessment of the relative importance of each user from the standpoint of the economy, environment, and national security. The matrices giving these judgments are respectively:

	Economy				Environment				National security		
	C_1	C_2	C_3		C_1	C_2	C_3		C_1	C_2	C_3
C_1	1	3	5		1	1/2	1/7		1	2	3
C_2	1/3	1	2		2	1	1/5		1/2	1	2
C_3	1/5	1/2	1		7	5	1		1/3	1/2	1

The corresponding normalized eigenvectors are respectively the three columns of the following matrix:

$$\begin{pmatrix} 0.64833 & 0.09382 & 0.53962 \\ 0.22965 & 0.16659 & 0.29696 \\ 0.12202 & 0.73959 & 0.16342 \end{pmatrix}$$

This matrix is multiplied by the vector $P(0)$ yielding the following vector which is already normalized, giving the eigenvector priorities of the activities C_1, C_2, and C_3:

$$\alpha \equiv \begin{pmatrix} \alpha_1 \\ \alpha_2 \\ \alpha_3 \end{pmatrix} = \begin{pmatrix} 0.55237 \\ 0.23606 \\ 0.21157 \end{pmatrix}$$

We cannot allocate energy in proportion to the priorities of the industries as they may be interdependent. Material from a low priority industry may flow to a higher priority one. To express this relationship as a constraint we use the following input-output matrix:

	C_1	C_2	C_3
C_1	1.09730	0.22680	0.19020
C_2	0.07990	1.06570	0.06010
C_3	0.03950	0.33210	1.20710

When the coefficient in the (i, j) position of the above matrix is weighted by α_i and α_j and summed over each row, we obtain the vector of dependence numbers:

$$\beta \equiv \begin{pmatrix} 0.38659 \\ 0.07280 \\ 0.07523 \end{pmatrix}$$

Suppose that the energy requirements R_i (in trillion BTU) of the three users are as follows:

Activity (C_i)	Energy requirements (R_i)
C_1	4616
C_2	7029
C_3	3297
Total	14942

Also assume that the total energy available has been cut back to a level of $R = 12,000$ BTU. We have the following linear programming problem:

Maximize

$$z = 0.38659w_1 + 0.07280w_2 + 0.07523w_3$$

whose coefficients are the corresponding elements of the vector β, subject to:

$$0 \le w_1 \le 0.38467, \quad 0 \le w_2 \le 0.58575, \quad 0 \le w_3 \le 0.27475,$$

in which the quantities on the right are respectively R_i/R, $i = 1, 2, 3$ and to

$$w_1 + w_2 + w_3 = 1.$$

The optimal allocation is given by

$$w_1 = 0.38467, \quad w_2 = 0.34058, \quad w_3 = 0.27475.$$

Thus only C_2 is not given its full requirement.

Note that here we have simplified the linear-programming problem to make it easier to grasp the procedure.

8.9.   Oil Prices in 1985

8.9.1.   THE PROBLEM

Today oil is the world's major energy resource. It accounts for about 54 % of the world's total energy consumption. Because of conservation and the development of alternative sources in the industrialized countries, the share of oil in the world's total energy consumption is expected to decline. But the total volume of oil consumption will still rise and it will remain the largest single source of energy supply for the next two decades (See Saaty and Gholamnezhad).

Despite oil price hikes between 1974 and 1979, the real price of OPEC oil has not increased significantly when adjusted to inflation and depreciation of the dollar. Actually, the devaluation of the dollar against the Japanese yen and the West German mark caused the real price of oil to decline in these countries. However, because of depletion of the world's proven oil reserves, increasing demand, and possible political unrest in the major oil-producing countries, oil prices are expected to rise during the next decade.

There has been a number of projections of world oil prices by major oil companies and government agencies. Most are based on demand and supply. But in today's world, oil-market economics and politics are interwoven and political decisions increasingly influence the levels of oil production, consumption, and prices.

We now use the analytic hierarchy process to project the real price of oil in 1985. Figure 8.3 represents the hierarchical model used.

8.9.2.   HOW TO COMPUTE PRICE INCREASES

1. Compute the relative weights of the factors $(W_1 \ldots W_5)$ according to their effectiveness in increasing the price of oil: W_1, world oil consumption increase; W_2, world excess production capacity; W_3, oil discovery rate; W_4, political factors; W_5, development of energy sources alternative to oil.

2. For each W_i, compute the relative likelihood of its corresponding subfactors. For $i = 1, 2, 3$ these are H_i, M_i, L_i (H_i, high; M_i, medium; L_i, low. See Fig. 8.3 for full details).

For example, for oil-consumption increase W_1 we ask the question: Which one of the three levels of increase is more likely for the period under consideration: 4%, 2%, or 1% per year? For $i = 4$, the subfactors are: P_1, instability of the Persian Gulf region; P_2, continuation of Arab-Israeli conflict; P_3, increasing Soviet influence in the Middle East and the relative importance of these is estimated. For $i = 5$, the subfactors are: V, vigorous; M, moderate; R, restrained; and the relative likelihood of those levels of development is estimated.

3. For the instability of the Persian Gulf region P_1 compute the relative importance of its three subfactors, namely: S_1, social strains within countries; S_2, tension between individual states; S_3, continuing disorder in Iran.

4. Compute the composite weights for each subfactors and select subfactors with high relative weights.

5. Compute the relative likelihood for each level of price increase for each selected subfactor.

6. Compute the composite weights of the levels of price increases. The result will be a set of numbers representing the likelihood of each price increase.

7. Compute the expected value of price increase by multiplying each price increase level by its corresponding likelihood. *Remark*: Some of the judgments used here were initially provided by five experts from major oil companies and were later modified to enforce consistency and to cope with other factors not included in the first version of this model. It may be useful for the reader to examine these judgments closely.

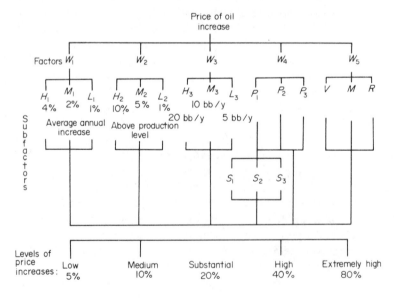

Fig. 8.3.

W_1	World oil-consumption increase	S_1	Social strains within countries
W_2	World excess production capacity	S_2	Tension between individual states
W_3	Oil discovery rate	S_3	Continuing disorder in Iran
W_4	Political factors	V	Vigorous
W_5	Development of energy sources alternative to oil	M	Moderate

P_1	Instability of the Persian Gulf region	R	Restrained
P_2	Continuation of Arab-Israeli conflict	H_i	High i = 1, 2, 3
P_3	Increasing Soviet influence in the Middle East	M_i	Medium
		L_i	Low
		bby	Billion barrels per year

8.9.3. COMPUTATION OF OIL PRICE IN 1985

M_1: Which factor would have a stronger effect on world oil prices by 1985?

Oil price 1985	W_1	W_2	W_3	W_4	W_5	Priorities
W_1	1	7	3	1/7	5	0.189
W_2	1/7	1	1/5	1/9	1/3	0.030
W_3	1/3	5	1	1/8	3	0.099
W_4	7	9	8	1	9	0.631
W_5	1/5	3	1/3	1/9	1	0.051

CR = 0.112

M_2–M_6: Which development is more likely to take place by 1985?

W_1	H_1	M_1	L_1	
4%/y H_1	1	3	7	0.649
2%/y M_1	1/3	1	5	0.279
1%/y L_1	1/7	1/5	1	0.072

CR = 0.056

W_2	H_2	M_2	L_2	
10% H_2	1	1/5	1/3	0.105
5% M_2	5	1	3	0.637
1% L_2	3	1/3	1	0.258

CR = 0.033

Oil-consumption increase:
$4 \times 0.649 + 2 \times 0.279 + 1 \times 0.072 = 3.2\%$ per year

Excess capacity:
$10 \times 0.105 + 5 \times 0.63 + 1 \times 0.258 = 4.5\%$ above production level in 1985

W_3	H_3	M_3	L_3	
20bb/y H_3	1	1/7	1/5	0.072
10bb/y M_3	7	1	3	0.649
5bb/y L_3	5	1/3	1	0.279

CR = 0.056

W_4	P_1	P_2	P_3	
P_1	1	3	5	0.637
P_2	1/3	1	3	0.258
P_3	1/5	1/3	1	0.105

CR = 0.033

Discovery rate:
$20 \times 0.072 + 10 \times 0.649 + 5 \times 0.279 = 9.3$ billion barrels per year.

M_6

W_5	V	M	R	
V	1	1/5	1/7	0.072
M	5	1	1/3	0.279
R	7	3	1	0.649

CR = 0.056

M_7: Which factor would have more effect on the instability of the Persian Gulf region by 1985?

	S_1	S_2	S_3	
S_1	1	3	5	0.637
S_2	1/3	1	3	0.258
S_3	1/5	1/3	1	0.105

CR = 0.033

TABLE 8.10

	H_1	M_1	L_1	H_2	M_2	L_2	H_3	M_3	L_3	S_1	S_2	S_3	P_2	P_3	V	M	R
Composite weights	0.123	0.053	0.014	0.003	0.019	0.008	0.007	0.064	0.028	0.256	0.104	0.042	0.163	0.066	0.004	0.014	0.033
Selected factors (normalized)	0.141			0.022			0.074			0.294	0.120	0.048	0.187	0.076			0.038

M8–M16: Given the following developments, which price-increase level is more probable?

H_1	E	H	S	M	L	
80% → E	1	1/3	1/5	1/3	1	0.071
40% → H	3	1	1/3	1	5	0.215
20% → S	5	3	1	3	5	0.460
10% → M	3	1	1/3	1	3	0.189
5% → L	1	1/5	1/5	1/3	1	0.065

CR = 0.025

M_2	E	H	S	M	L	
E	1	1/3	1/5	1/7	3	0.063
H	3	1	1/3	1/5	5	0.129
S	5	3	1	1/3	7	0.261
M	7	5	3	1	9	0.513
L	1/3	1/5	1/7	1/9	1	0.033

CR = 0.053

M_3	E	H	S	M	L	
E	1	1/4	1/5	1/3	2	0.070
H	4	1	1/2	2	5	0.243
S	5	2	1	7	6	0.496
M	3	1/2	1/7	1	5	0.146
L	1/2	1/5	1/6	1/5	1	0.045

CR = 0.067

S_1	E	H	S	M	L	
E	1	1/5	1/3	1	3	0.101
H	5	1	3	5	7	0.502
S	3	1/3	1	3	6	0.251
M	1	1/5	1/3	1	3	0.101
L	1/3	1/7	1/6	1/3	1	0.044

CR = 0.028

S_2	E	H	S	M	L	
E	1	1/3	1/5	1/7	1/3	0.046
H	3	1	1/3	1/5	1	0.102
S	5	3	1	1/3	3	0.245
M	7	5	3	1	5	0.504
L	3	1	1/3	1/5	1	0.102

CR = 0.028

S_3	E	H	S	M	L	
E	1	1/3	1/5	1/3	1	0.073
H	3	1	1/3	1	3	0.194
S	5	3	1	3	5	0.466
M	3	1	1/3	1	3	0.194
L	1	1/3	1/5	1/3	1	0.073

CR = 0.012

P_2	E	H	S	M	L	
E	1	1/3	1/6	1/3	1	0.066
H	3	1	1/4	1	3	0.171
S	6	4	1	4	6	0.528
M	3	1	1/4	1	3	0.171
L	1	1/3	1/6	1/3	1	0.066

CR = 0.017

P_3	E	H	S	M	L	
E	1	1/3	1/5	1	3	0.102
H	3	1	1/3	3	5	0.245
S	5	3	1	5	7	0.504
M	1	1/3	1/5	1	3	0.102
L	3	1/5	1/7	1/3	1	0.046

CR ≈ 0.028

R	E	H	S	M	L	
E	1	1/3	1/5	3	5	0.129
H	3	1	1/3	5	7	0.261
S	5	3	1	7	9	0.513
M	1/3	1/5	1/7	1	3	0.063
L	1/5	1/7	1/9	1/3	1	0.033

CR = 0.053

8.9.4. ANALYSIS OF THE RESULTS FOR 1985

We now analyze the factors influencing world oil prices according to their relative importance to show how the comparisons were made.

a. *Political Factors* ($W_3 = 0.631$)

Political factors play an extremely important role in the world oil market. The Arab oil embargo of 1973, the Iranian revolution, and consequent disruptions in the world oil supplies have demonstrated the significance of political factors in the supply, demand, and price of oil.

Political factors included in this analysis are the instability of the Persian Gulf region, continuation of the Arab-Israeli conflict and increasing Soviet influence in the Middle East. Although OPEC itself plays an important political role in the oil market, its stability is very dependent on developments in the Middle East.

P_1: *Instability in the Persian Gulf region* (0.402). The region which will continue to be of extreme importance in the future supply and prices of oil is the Middle East, particularly the Persian Gulf states. The Persian Gulf is surrounded by a number of major oil-exporting countries such as Iran, Saudi Arabia, Iraq, Kuwait, Qutar, Bahrain, and the United Arab Emirates. These countries, excluding Bahrain, which is not a major oil exporter, are the members of OPEC and altogether account for over 80 % of its proved oil reserves or nearly half of the world's total reserve. Over 30 % of the world's oil supply comes from this region. Stability of the Persian Gulf itself depends on several other factors, particularly the social strains due to rapid economic developments, industrialization, unstable political systems, and religious movements in the region. Also, tensions between the individual states, as we are now witnessing between Iran and Iraq, could lead to a regional war. Another factor to be considered is the continuation of disorder in Iran which would not only keep its oil output as low as it is today, but may also increase instability in the region.

P_2: *Continuation of the Arab-Israeli conflict* (0.163). The Arab oil embargo of 1973 demonstrated the impact of the Arab-Israeli conflict on the flow of oil to the industrialized world. Long delays in peace will discourage the major Arab oil producers from cooperating in meeting the demands of the industrialized world. This will put more pressure on the world oil market and consequently oil prices will rise drastically.

P_3: *Increasing Soviet influence in the Middle East* (0.066). Although the Soviet Bloc is currently a net exporter of oil, because of decline in oil production in the Soviet Union, it is projected to become a net importer of oil in the near future. Therefore, the Soviet Union will be competing with Western countries for Middle Eastern oil. Some political observers believe that the purposes of Soviet intervention in Afghanistan and its assumed assistance to the rebels in Baluchistan include providing itself with secure oil and gas supply sources in the future by assured access to the Persian Gulf. Increasing Soviet influence in the Middle East will enhance its position in the oil market *vis-à-vis* the West. And, if it becomes to their advantage, the Soviets would not hesitate to use oil as an economic weapon against the West, particularly the United States. This action will lead to higher payments for oil for the Western countries.

b. *World oil-consumption Increase* ($W_1 = 0.123$)

In 1979, the United States, Japan, and Europe accounted for about 75 % of the total world oil consumption. No substantial increase in demand is anticipated for these countries but, in the developing countries, particularly the oil-exporting countries, demand for oil is expected to increase significantly due to industrialization and development.

c. *Oil Discovery Rate* ($W_3 = 0.099$)

Before 1970 oil discovery rates were much higher than oil-production rates. Therefore, the volume of the world's discovered reserves was increasing. But, since the early 1970s, oil discovery declined steadily while production rates increased continuously. This downward trend for discovery rates is projected to continue slowly until 1985 and rather sharply thereafter.

d. *Development of Energy Sources Alternative to Oil* ($W_5 = 0.051$)

A substantial amount of oil could be replaced by synthetic fuels from large coal, oil shale, tar sands reserves, and biomass resources but, due to the long lead times (about 6 to 10 years), large capital requirements and environmental constraints, such fuels are not expected to make a significant contribution during the next decade. However, in the 1990s synfuels will play an important role in the world energy market.

e. *World Excess Production Capacity* ($W_2 = 0.030$)

Today, the world's excess production capacity is more than 10 million barrels per day (MBD), two-thirds of which is from the Middle East. At this level of excess capacity only large oil producers can affect oil prices by making their production levels fluctuate. However, when the excess capacity declines substantially, say to 2 to 3 MBD, even small producers can cause a sudden jump in oil prices by cutting back their production (or large producers by cutting back on a small portion of their production).

f. *Projected Oil-Price Increase* (1985)

Table 8.11 shows the probabilities of given oil-price increase for each level under consideration. According to these results, the price increase for 1985 would be:

TABLE 8.11. *Probabilities of given levels of price increase by 1985*

Level	%	Composite probability
E	80	0.080
H	40	0.281
S	20	0.389
M	10	0.190
L	5	0.059

$$80 \times 0.080 + 0.40 \times 0.281 + 20 \times 0.389 + 10 \times 0.90 + 5 \times 0.059 = 27.6\%$$

given the present price of Arabian light crude (market crude) of $32 per barrel, a 27.6% increase by 1985 means that the *real* price of oil will be

$$32 + 32 \times 0.276 = \$40.80.$$

Assuming an average inflation rate of 10% in the United States, Americans will pay 40.80 $(1 + 0.10)^5 = \$65.70$ (or more, depending on the quality of crude oil) for each barrel of imported oil by 1985.

Our results are higher than those projected by the Exxon Corporation (World Energy Outlook, 1980). Based on the price of Arab oil (light crude) in October 1979 of $18 per barrel, the Exxon study projected that the real price of oil would be $25 per barrel in 1985 and $28 per barrel in 1990. The Exxon results are below those assumed by the US Senate Finance Committee in its projections.

8.10.   Architectural Design

In our final example, we combine the analytic hierarchy process with other considerations, such as aesthetic qualities (see Saaty and Beltran).

We want to design a house for a family of three (husband, wife, and child) who own a plot on which they want to build a custom-designed house. Their maximum budget is $105,000 to cover construction and landscaping but not legal and architectural fees, etc. The cost of construction is assured to be $45 per square foot and that of landscaping is $1.80 per square foot.

The members of the family are the decision-makers. We assume that they will seek to satisfy the same human needs, but may assign different weights to each objective. The objectives are (i) the need to eat (M), (ii) the need to rest (R), (iii) the need to entertain (including each other) (E), (iv) the need to clean (people, clothes, etc.) (K), (v) the need to store (S), and (vi) the need to communicate within the house (by hallways, etc.) (C). Certain rooms meet a number of these objectives: for example, the family room serves for resting, eating, and entertaining.

This gives us initially a hierarchy for allocation of space within the house according to the way in which each room satisfies the needs of family members. This hierarchy is shown in Fig. 8.4 and the abbreviations used in the judgmental matrices are taken from this diagram.

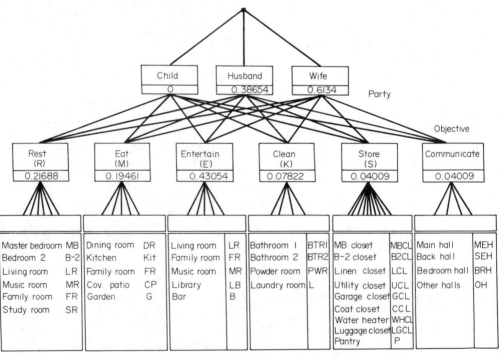

Architectural spaces

Fig. 8.4.

Our first calculation shows that the child will have little influence on the final outcome, even if he forms a coalition with his father, and so we consider only the needs of the husband and wife. The particular husband and wife in this situation agreed on the judgments as given; they would not necessarily be true for all families (Table 8.12).

TABLE 8.12. *Power of parties—husband, wife, child*

	W	H	CH	Weights	Revised Power
W	1	2	4	0.5469	0.6134
H	0.5	1	4	0.3445	0.3866
CH	0.25	0.25	1	0.1085	0

As before, we must now prioritize the importance of the objectives from the standpoint of both husband and wife, and we also include combined weights. This leads to Table 8.13.

TABLE 8.13 (*a*) *Strength of objectives*

	R	M	E	K	S	C	Eigenvector	Power of party	Objectives adjusted for power
Party: Wife									
R	1	0.5	0.33	4	6	6	0.1979		0.12140
M	2	1	0.33	3	4	4	0.2064	0.61346	0.12662
E	3	3	1	6	7	7	0.4244		0.26035
K	0.25	0.33	0.16	1	4	5	0.09637		0.05912
S	0.16	0.25	0.14	0.25	1	1	0.03783		0.02321
C	0.16	0.25	0.14	0.20	1	1	0.03708		0.02321
							Lamba: 6.456		
							CI: 0.09		
Party: Husband									
R	1	2	0.33	5	6	6	0.247		0.09548
M	0.5	1	0.33	4	5	5	0.1759		0.06799
E	3	3	1	6	8	8	0.4403	0.38654	0.17019
K	0.2	0.25	0.16	1	1	1	0.04942		0.01910
S	0.16	0.20	0.125	1	1	1	0.04366		0.01688
C	0.16	0.20	0.125	1	1	1	0.04366		0.01688
							Lambda: 6.141		
							CI: 0.03		

(*b*) *Combined weight of objectives*

	Wife	Husband	Combined weight
R	0.12140	0.09548	0.21688
M	0.12662	0.06799	0.19461
E	0.26035	0.17019	0.43054
K	0.05912	0.01910	0.07822
S	0.02321	0.01688	0.04009
C	0.02321	0.01688	0.04009

We then allocate the space available for each activity from the composite outcomes. The family have decided, on the advice of an architect, to allocate about 85 % or about $90,000 to construction and the remainder to landscaping. At $45 per square foot, this means that the house would have a maximum area of 2000 square feet. This leads to the following allocation:

Total area to allocate: 2000 sq. ft.

Objective	R	M	E	K	S	C
Strength of Objective	0.2169	0.1946	0.4305	0.0782	0.0401	0.0401
Area per objective	433.8	389.2	861.0	156.4	80.2	80.2

We then allocate the space in each category according to the relative importance given to each room by the parties. Since each views this differently, this must also be weighted according to their relative power to obtain a fair division of space. Note that some rooms will draw space for more than one category.

We give here in Table 8.14 the resulting allocation. (The reader is referred to Saaty and Beltran for details of judgmental matrices.)

TABLE 8.14. *Allocation of areas to architectural spaces*

Objective	Area of objective (sq. ft.)	Architectural space	Party's preferences W	H	Combined weight	Allocated area to space.
R	433.70	MB	0.41918	0.2449	0.66413	288.00
		B-2	0.07165	0.0674	0.13906	60.31
		SR	0.12257	0.0741	0.19674	85.33
M	389.20	DR	0.36949	0.2121	0.58162	226.36
		FR	0.06822	0.0746	0.14282	55.58
		KIT	0.13870	0.0702	0.20897	81.33
		CP	0.03706	0.0295	0.06656	25.90
E	861.00	LR	0.32035	0.2018	0.52220	449.61
		FR	0.10680	0.0672	0.17409	149.89
		MR	0.05994	0.03776	0.09770	84.12
		LB	0.10680	0.06729	0.17409	149.89
		B	0.01951	0.01229	0.03180	27.38
K	156.40	BTR1	0.34955	0.22025	0.56980	89.12
		BTR2	0.15263	0.09617	0.24880	38.91
		PWR	0.03540	0.02230	0.05770	9.02
		L	0.07589	0.04781	0.12370	19.34
S	80.00	MBCL	0.33673	0.21217	0.54890	43.91
		B2CL	0.14883	0.09377	0.24260	19.40
		P	0.07220	0.04549	0.11769	9.41
		CCL	0.02411	0.01521	0.03932	3.14
		GCL	0.03153	0.01988	0.05141	4.11
C	80.00	MEH	0.07397	0.04661	0.12058	9.64
		SEH	0.02711	0.01708	0.04419	3.53
		BRH	0.16709	0.10529	0.27238	21.79
		OH	0.34516	0.21750	0.56266	45.01

We then have to decide on appropriate dimensions and this may be done by selecting varying sets of dimensions and prioritize: some adjustments may have to be made to maintain the totals for each objective. A final set of dimensions may then be obtained. In our example, the total area was 1982 sq. ft., which does not exceed the maximum permissible.

It remains to locate the different rooms. We focus on the three most important areas: entertainment, rest, and meals, and place these first:

	Weights	Adjusted weights
E	0.4305	0.5113
R	0.2169	0.2576
M	0.1946	0.3211
	0.8420	1.0000

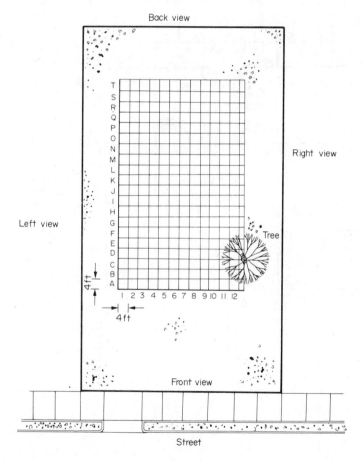

FIG. 8.5.

The criteria used are front view FV, back view BV, left-side view LV, and right-side view RV (with a tree). The husband and wife again decide how they wish each area to be located by function. They then decide how to place the highest priority room, the living-room, in the highest priority location for entertainment needs, the front view. The house is divided into quadrants, as shown in Fig. 8.5 and blocks are prioritized. The quadrant $C_{6-10}H_{6-10}$ received the highest weight and was selected. We position the remaining entertainment rooms and rooms in other areas similarly, in descending orders of priority. The remaining regions (storage, etc.) are then fitted in. The result is shown in Fig. 8.6.

We have thus formalized a singularly complex decision process by using a number of simpler decisions and, by obtaining input at each stage, have combined both intuitive and more formal information in the same model.

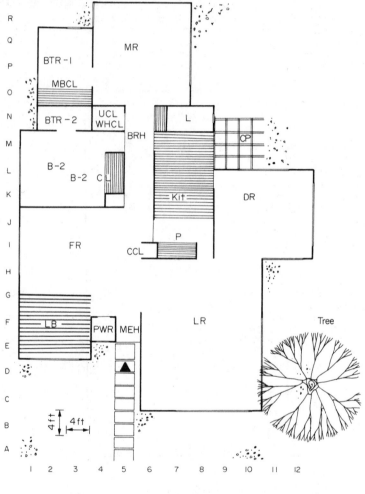

FIG. 8.6.

Chapter 8—Problems

1. Describe some decision which you have to make in the near future. Formulate an analytic hierarchy to help you make this decision and find the "best" outcome. Does this outcome surprise you? Why? Now enlarge the problem to include other people who may influence the final outcome. Does the answer change? Why?

2. You have been asked to advise a company on the allocation of its resources between the development costs of two new products. Formulate an analytic hierarchy to solve this problem.

3. A large insurance company has decided to diversify its portfolio. Formulate an analytic hierarchy to decide among a range of possible purchases, taking into account both the internal characteristics of the firm and the external factors of the market.

4. You are entering the job market shortly. What factors are important in choosing a job? Formulate a decision model to help you in your choice.

5. A family is deciding on a city in which to live. Develop a hierarchy to assist in the decision, considering physical, economic, and social factors.

6. A manufacturing company has decided to open a plant overseas, but has not yet decided on the country. Formulate a hierarchy to assist in this decision. Remember to take into account such factors as managerial and operational control, tariffs, capital formation, labor conditions, social and cultural factors, and political risks.

7. Using an appropriate hierarchy determine the expected cost of a compact automobile by 1990 allowing for inflation. It is the object of this exercise that one include the so called imponderables.

8. The Saaty family considered buying a home computer, but the computer was not the only source of pressure on their resources at the time. At the time the kitchen was being remodelled causing considerable financial drain. Therefore even if the best computer were determined, it would have contended with saving the money in the bank, cementing the driveway or even donating the money to charity. Construct a hierarchy of benefits and a hierarchy of costs (not simply dollar costs but other sources of pain) to decide which of these computers, Radio Shack, Apple II, and Texas Instruments would have the highest benefit to cost ratio. Then take the best one along with keeping the money in the bank, and performing other urgent large expenditures to decide on the benefit to cost ratio. Note that the results may be suggesting an order of implementation over time.

References

Alexander, J. M. and T. L. Saaty, The forward and backward processes of conflict analysis, *Behav. Sci.*, Vol. 22, 1977, pp. 87–98.

Alexander, J. M., and T. L. Saaty, Stability analysis of the forward–backward process: Northern Ireland case study, *Behav. Sci.*, Vol. 22, 1977, pp. 375–382.

Miller, George A., The magical number seven, plus or minus two: some limits on our capacity for processing information, *Psychological Rev.*, Vol. 63, March 1956, pp. 81–97.

Saaty, T. L., *The Analytic Hierarchy Process*, McGraw-Hill, New York, 1980.

Saaty, T. L., A scaling method for priorities in hierarchical structures, *J. Math. Psychol.*, Vol. 15, 1977, pp. 234–281.

Saaty, T. L., The Sudan transport study, *Interfaces*, Vol. 8, No. 1, 1977a, pp. 37–57.

Saaty, T. L., and R. S. Mariano, Rationing energy to industries: priorities and input–output dependence, *Energy Systems and Policy*, Vol. 8, 1979, pp. 85–111.

Saaty, T. L., Scenarios and priorities in transport planning: application to the Sudan, *Transport. Res.*, 1977b.

Saaty, T. L., and M. Khouja, A measure of world influence, *J. Peace Science*, Vol. 2. No. 1, 1976, pp. 31–47.

Saaty, T. L., and P. C. Rogers, The future of higher education in the United States, *Socio-econ. Plan. Sci.*, Vol. 10, No. 6, 1976, pp. 251–264.

Saaty, T. L., and M. Beltran, Architectural design by the analytic hierarchy process, *J. Design Methods Group*, June, 1980.

Saaty, T. L., and A. H. Gholamnezhad, Estimates of oil prices in 1985 and 1990 (to appear).

Wilkinson, J. H., *The Algebraic Eigenvalue Problem*, Oxford, Clarendon Press, 1965.

Index